Benchmarking Water Services
Guiding water utilities to excellence

Benchmarking Water Services
Guiding water utilities to excellence

Enrique Cabrera Jr., Peter Dane,
Scott Haskins and Heimo Theuretzbacher-Fritz

Published by   IWA Publishing
                 Republic - Export Building
                 1 Clove Crescent
                 London E14 2BA, UK
                 Telephone: +44 (0)20 7654 5500
                 Fax: +44 (0)20 7654 5555
                 Email: publications@iwap.co.uk
                 Web: www.iwaponline.com

First published 2011
© 2011 IWA Publishing

Cover design by www.sixteen-design.co.uk
Typeset by MPS Limited, a Macmillan Company, Chennai, India.

Disclaimer
The information provided and the opinions given in this publication are not necessarily those of IWA and should not be acted upon without independent consideration and professional advice. IWA and the Author will not accept responsibility for any loss or damage suffered by any person acting or refraining from acting upon any material contained in this publication.

British Library Cataloguing in Publication Data
A CIP catalogue record for this book is available from the British Library

Library of Congress Cataloging-in-Publication Data
A catalog record for this book is available from the Library of Congress

ISBN: 9781843391982 (Hardback)
ISBN: 9781789065190 (Paperback)
ISBN: 9781780400877 (Ebook)

Contents

Chapter 6
Orientation, training and project control 57

Chapter 7
Data acquisition and validation................................65

Chapter 8
Data analysis & assessment reporting.......................73

Chapter 9
Improvement actions 95

Chapter 10
Review improvement actions 105

Foreword

In 1996, the AWWA Research Foundation published a manual on benchmarking that ignited great interest in the industry. Approximately at the same time, a benchmarking task force was being created at IWA's Statistics and Economics Specialist Group. Those were the early days of interest in performance assessment and benchmarking, an interest that was probably born from the seed planted by sector regulation in England and Wales in the late 90s. The first signs that this topic was relevant enough to a significant number of players appeared in 2000, after the publication of the IWA manual of best practice on performance indicators for water supply.

During the past decade, the water industry witnessed many initiatives that helped to push forward knowledge on performance assessment and benchmarking in the industry. Some of those efforts are referenced in Annex A and other parts of this manual, but the task of acknowledging all those pioneers certainly laid beyond our reach. Many of those initial efforts have either faded away or evolved into completely renewed projects, but it was certainly through those efforts that many of the industry players quietly began considering performance indicators and benchmarking as a real tool that could be applied in the real world.

As knowledge evolved in real world projects, so did its gathering in many groups and organizations. IWA published in 2002 a manual of best practice on process benchmarking, which was to establish a mangement tool to allow for performance comparisons that could lead to improvements in performance and quality. In 2003, the performance indicator system for wastewater services was published, and in 2006, the second edition of the manual for water supply saw the light including the complete definition of IWA's performance assessment framework. Meanwhile, AWWA published in 2005 and 2006 two reports on performance indicators and benchmarking for water and wastewater utilities.

The manual in your hands follows up and is consistent with all those previous efforts. However, if a difference must be made between this text and those that preceeded it, this would probably be its more applied nature. The water industry has been benchmarking quite intensely in the past decade. There are several consolidated initiatives around the world from which a great deal can be learnt, and we certainly tried to include those lessons in the manual.

However, the manual also intends to provide a common framework and language for the benchmarking community. During all these years, we have often found that words stood in the way of communication between water professionals on the topic. Terms like process, metric and even benchmarking became sometimes more important than the useful information that could be obtained from a certain initiative. This manual presents a new framework, hoping that water professionals around the world will embrace it and use it in the future. The fact that the manual is co-published by the two largest and most influential organizations in the industry (IWA and AWWA) is certainly a huge step in that direction.

Benchmarking is a key topic for many of the stakeholders in the industry. From administrations, regulators, researchers, funding agencies and consumers to the obvious inclusion of utilities and their associations, many actors have shown interest on the topic. However, the information contained in the manual has been mainly addressed to the facilitators of the benchmarking projects and the participating utilities, with all texts inside a shaded box specifically addressed to water utilities. Needless to say, that regardless of the reader's background, we think that the manual contains valuable information for anyone interested in benchmarking in the water sector.

A long journey, spanning nearly a decade, finally reaches its destination with the publication of this manual. But far from being the end, a new journey simultaneously begins, for the new IWA Specialist Group on Benchmarking and Performance Assessment has recently been created. An open forum in which to share experiences and learn from others, which after all is the true spirit of benchmarking.

The authors.

Acknowledgments

The greatest risk in including a note acknowledging who made this manual possible is to leave someone out. This book has been possible through the work, input and collaboration of many people, and we would like to apologize in advance in case we cannot manage to include them all in this brief note.

This manual (as so many others) originated in a voluntary project that started in 2002. That project was possible through a combined effort of two IWA specialist groups. Francisco Cubillo and Renato Parena deserve the credit for making it happen.

During that time, some people worked at the heart of the project and contributed to the final result. Peter Gee from WSAA helped us a lot along the way and many of his ideas and work have made it into the book, even if his name is not on the cover. Jens Bastrup, Ingrid Troquet and Ed Smeets represented at different times the Statistics and Economics SG in the taskforce, therefore contributing to the final result.

Emma Rose and Keith Robertson from IWA sat in all of our meetings and had to suffer the theoretical and practical discussions leading to this manual. Their contribution was vital in getting the manual finished.

IWA partially funded the project by covering some travel expenses for our meetings. The employers of the authors financed the rest, allocating both time and resources to the project.

Several associations with benchmarking projects were kind enough to provide us with valuable information and documents for Annex B. The European Benchmarking Cooperation (EBC), the Water Services Association of Australia

© 2011 IWA Publishing. *Benchmarking Water Services: Guiding water utilities to excellence.* By Enrique Cabrera Jr., Peter Dane, Scott Haskins and Heimo Theuretzbacher-Fritz. ISBN: 9781789065190. Published by IWA Publishing, London, UK.

(WSAA), the Austrian Association for Gas and Water (OVGW) and the Association of Dutch Water Companies (VEWIN). All the utilities participating in their benchmarking programmes are duly acknowledged as well. Special mention should be made to the utilities of Anchorage, Seattle and Tacoma for their contribution to the annexes. Several people around the world directly or indirectly contributed with information about the case studies in Annex A.

Finally, we would like to thank IWA Publishing and AWWA's publishing program for providing the largest possible showcase within the water industry for this manual.

Enrique Cabrera & Heimo Theuretzbacher-Fritz

About the authors

Enrique Cabrera is Associate Professor of fluid mechanics at the Universidad Politecnica de Valencia, Spain. Also a co-author of the IWA Manual of Best Practice on Performance Indicators for Water Supply Services, Enrique has been working for well over a decade on the topic of performance assessment and benchmarking.

Enrique has played an active role in IWA, where he acted as secretary of the Efficient Operation and Management of Urban Water Systems and Planning and Construction Specialist Groups. In 2008, he received the Association's Young Water Professional Award. Currently he co-chairs the IWA Specialist Group on Benchmarking and Performance Assessment.

His international experience includes his role as a convenor of one of the working groups drafting of the ISO 24500 standards on the assessment and improvement of water services. He has also acted on advisory and consulting roles in performance assessment projects worldwide, and published several books and numerous papers and journal articles on the topic.

Enrique holds two master degrees in industrial engineering and efficient use and management of water, and a PhD with a thesis on Performance Assessment and Benchmarking of water services by the Universidad Politecnica de Valencia.

Peter Dane is manager of international benchmarking at Vewin, the Association of Dutch Water companies, the Netherlands. In this position, he manages the European Benchmarking Co-operation, a joint benchmarking initiative of Dutch and Scandinavian partners. In 1981 he joined the municipal water supply company of the city of Rotterdam, where he was head of the Department for corporate planning, marketing and industrial water services between 1986 and 1992. From 1989–1994 he was member of the core team that merged ten water utilities in the greater Rotterdam region. Between 1992–2004 he was manager of Corporate Strategy, responsible for business planning, legal advices and R&D. From 1993–1995, he was member of the National Committee that developed a guideline for reliability of water supply systems. Between 2002 to 2004 he joined the National Committee that advised on measures to secure water supply.

Peter is member of the Benchmarking Task Group of IWA. He frequently gives presentations at international workshops and conferences and is co-author of the book "Private business, public owners", describing the Dutch model of water supply.

Peter holds a MSc in civil engineering from Delft University of Technology and a BSc in Business Administration from the Dordrecht School for Business Administration.

Scott Haskins is Vice President and Director for Technology, Quality and Innovation for CH2M HILL. In his consultant role, he coordinated North American efforts for WSAA/IWA's Asset Management Benchmarking Project, and is also directing other asset management and management consulting projects.

Previously, he concluded a career with the City of Seattle as Deputy Director for Seattle Public Utilities where he was responsible for drinking water, surface water, wastewater and solid waste functions. Scott has been a very active participant in the water/wastewater utility industry, having written many papers and journal articles, given frequent presentations at national and international workshops and conferences, and participated in numerous research roles and projects for national and international organizations. He is co-author of the AWWA books, The Changing Water Utility; Creative Approaches to Effectiveness and Efficiency, and The Evolving Water Utility: Pathways to High Performance.

Scott holds a Master of Public Administration and a Bachelor of Arts Degree in Political Science from the University of Washington.

Heimo Theuretzbacher-Fritz is senior researcher at Graz University of Technology in Austria, at the Institute of Urban Water Management and Water Landscape Engineering. He has been working in the applied research area of water supply, with special emphasis on strategic utility management, on benchmarking and performance assessment, and on quality and risk management.

Heimo is a leading member of the Austrian water supply benchmarking team, commissioned by OVGW, the Austrian Association of Gas and Water, and conducting benchmarking activities both on utility and process level since 2003.

Since 2005, Heimo has also engaged in benchmarking efforts at an international level (Bavaria, Slovenia, European Benchmarking Co-operation). Furthermore, he is chairing the Task Force on Process Benchmarking of the IWA Specialist Group on Statistics & Economics and is co-chairing the new IWA Specialist Group on Benchmarking and Performance Assessment.

Heimo holds a Master of Science Degree from Graz University (curriculum: Environmental Systems Science). His PhD thesis "Strategic Management of Water Supply Utilities" within the doctoral school of Civil Engineering Sciences at Graz University of Technology is currently in the final stage of preparation. Heimo is author or co-author of 50 publications and 79 project reports and expertises, and gave 38 presentations (half of them on international events).

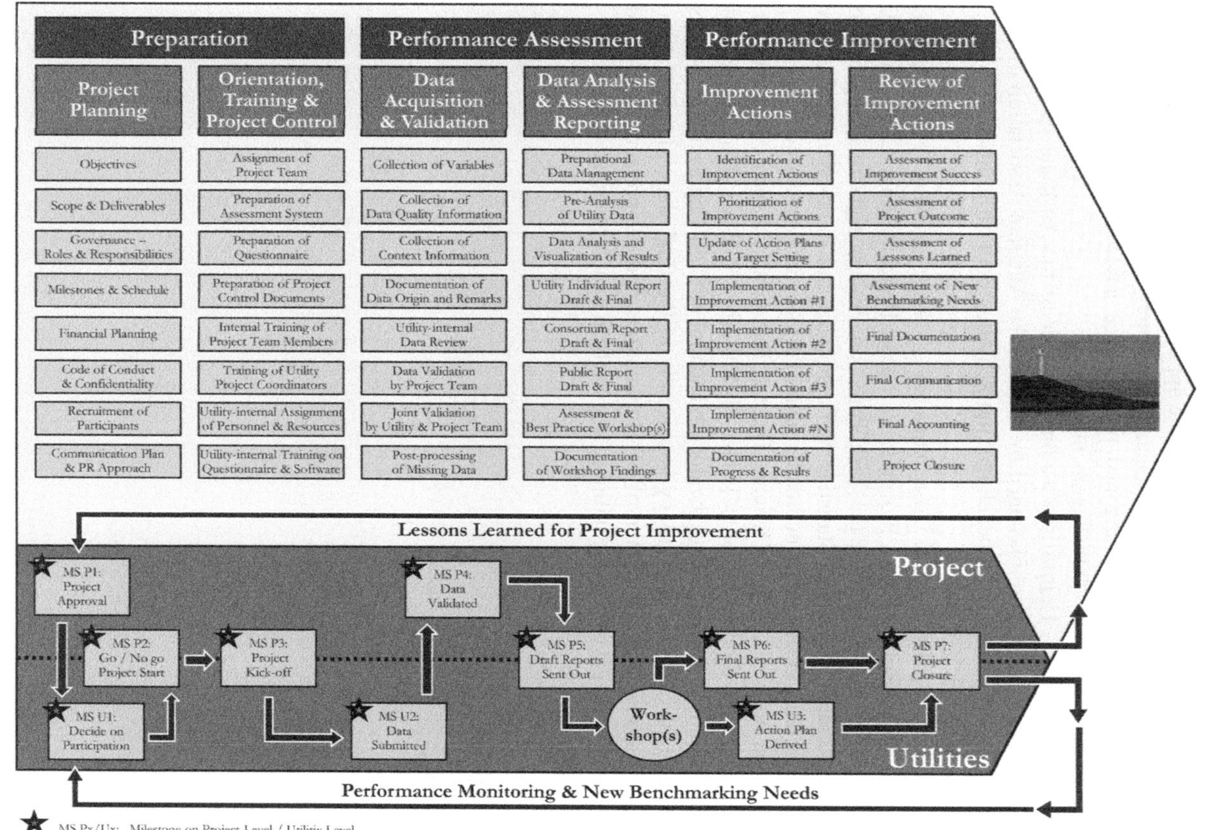

The IWA Benchmarking Framework

Preparation
Performance Assessment
Performance Improvement

Project Planning
Orientation, Training & Project Control
Data Acquisition & Validation
Data Analysis & Assessment Reporting
Improvement Actions
Review of Improvement Actions

Objectives
Assignment of Project Team
Collection of Variables
Preparational Data Management
Identification of Improvement Actions
Assessment of Improvement Success

Scope & Deliverables
Preparation of Assessment System
Collection of Data Quality Information
Pre-Analysis of Utility Data
Prioritization of Improvement Actions
Assessment of Project Outcome

Governance – Roles & Responsibilities
Preparation of Questionnaire
Collection of Context Information
Data Analysis and Visualization of Results
Update of Action Plans and Target Setting
Assessment of Lessons Learned

Milestones & Schedule
Preparation of Project Control Documents
Documentation of Data Origin and Remarks
Utility Individual Report Draft & Final
Implementation of Improvement Action #1
Assessment of New Benchmarking Needs

Financial Planning
Internal Training of Project Team Members
Utility-internal Data Review
Consortium Report Draft & Final
Implementation of Improvement Action #2
Final Documentation

Code of Conduct & Confidentiality
Training of Utility Project Coordinators
Data Validation by Project Team
Public Report Draft & Final
Implementation of Improvement Action #3
Final Communication

Recruitment of Participants
Utility-internal Assignment of Personnel & Resources
Joint Validation by Utility & Project Team
Assessment & Best Practice Workshop(s)
Implementation of Improvement Action #N
Final Accounting

Communication Plan & PR Approach
Utility-internal Training on Questionnaire & Software
Post-processing of Missing Data
Documentation of Workshop Findings
Documentation of Progress & Results
Project Closure

Lessons Learned for Project Improvement

Project

MS P1: Project Approval
MS P2: Go / No go Project Start
MS P3: Project Kick-off
MS P4: Data Validated
MS P5: Draft Reports Sent Out
MS P6: Final Reports Sent Out
MS P7: Project Closure

MS U1: Decide on Participation
MS U2: Data Submitted
Work-shop(s)
MS U3: Action Plan Derived

Utilities

Performance Monitoring & New Benchmarking Needs

MS Px/Ux: Milestone on Project Level / Utilitiy Level

Chapter 1
Introduction

This manual aims to provide a guide for water utilities on why and how to benchmark. And while presenting the key concepts and clarifying some of the industry's misconceptions is inevitable, a great effort has been made to provide a practical reference with examples and applicable knowledge obtained first hand from some of the key benchmarking initiatives in the water industry around the world.

The text reflects the work of the IWA benchmarking task group, which was created with the purpose of continuing on previous works on performance indicators and benchmarking, and producing this manual to facilitate the understanding of all those benchmarking worldwide, and provide valuable knowledge to those wanting to benchmark.

1.1 WHAT IS BENCHMARKING?

Most benchmarking books written in the past twenty years have tried to answer this question on their first pages. After all benchmarking is a complex concept and a good book should try to interest the reader from the beginning laying out the topic clearly and concisely.

This has resulted in a handful of different ways to describe benchmarking. However, the truth is that with very few exceptions, those books had an easier task than this one you are holding right now, for they were rarely addressed to the water industry. This should not be taken to mean that benchmarking is any different when applied to a water utility than to a manufacturing company. It is

© 2011 IWA Publishing. *Benchmarking Water Services: Guiding water utilities to excellence.* By Enrique Cabrera Jr., Peter Dane, Scott Haskins and Heimo Theuretzbacher-Fritz. ISBN: 9781789065190. Published by IWA Publishing, London, UK.

not. But after a decade and a half of sometimes confusing terminology, water professionals may have a harder time with benchmarking concepts than the average reader (Cabrera *et al.*, 2009a).

One of the main aims of this manual is to try to provide clear definitions and a single benchmarking language for the water industry. A task posing a challenge even greater than describing benchmarking in a single phrase:

> **"Benchmarking is a tool[1] for performance improvement through systematic search and adaptation of leading practices"**

The careful choice of words aims to capture in a single sentence a broad concept. And key to that concept is the idea that benchmarking is simply a tool. A powerful tool that is especially suited for the water industry, but in no way an end in itself. Benchmarking without a clear goal will often lead to disappointment and waste of resources.

Another cornerstone of the benchmarking concept is captured by the word systematic. Benchmarking techniques should always be aimed at continuous improvement. As a matter of fact, benchmarking fits especially well in the PDCA (Plan, Do, Check, Act) concept and should always be approached with those four stages in mind. As a natural consequence, benchmarking should derive in any organization in a natural tendency towards continuous improvement.

However benchmarking is anything but a "do it yourself" concept. The search for leading practices implies that lessons need to be learnt from others[2], hopefully from the best in class. But as much as benchmarking is about looking outside, it is also an exercise of looking within and learning how things are done internally. It is only from that inner knowledge and the understanding of how others (the best) do things, that improvement can be achieved.

An obvious factor to that success is understanding the concepts of *best practices* and *best in class*. Those words are often quoted in the literature pointing to iconic case studies and world-class companies as references in certain processes. While it is true that those organizations constitute a clear reference, benchmarking is not limited to industry leaders and large organizations. And as a matter of fact, some of the benchmarking projects undertaken lately by water utilities show that lessons can be learnt from almost

---

[1] Benchmarking should probably be considered more a process than a tool. However, for reasons later explained, the authors have refrained from using the term "process" in this definition.

[2] These others may certainly include other utilities from a same group of companies.

anyone, and that being able to identify the best practices and those who have developed them is a key factor to success.

The truth is that to find out whom the best in class is, measurements need to be made. However, and despite the very extended notion that benchmarking, or some type of benchmarking, is achieved by comparing metrics, it should be made very clear that creating a bar graph with a numerical comparison of several utilities is not benchmarking. Even the original definition of "metric benchmarking" (which will be reviewed later in this chapter) considered it to be much more than the comparison of a few indicators. The benchmarking concept has always encompassed a systematic process and the will for continuous improvement.

The consequences of continuous benchmarking efforts in a utility are often reflected in a more mature organization which is more transparent and understands better how things are done, which things need to and could be improved and what it takes to improve them.

1.2 BENCHMARKING: METRIC, PROCESS OR NONE?

Even from the early days, the water industry was quite fast in reacting to a new management tool called benchmarking. It was the early 90s, and the echoes from Harvard had already been made available to the general public in what many considered the first mainstream publication on the topic (Camp, 1989). From the very beginning, benchmarking was seen as a tool that, in the absence of efficiency stimulating incentives, could help water companies in their search for excellence.

The aim of benchmarking was to identify those who were best in their class, and once the best practices had been identified, they were to be adapted in search of better performance. Robert Camp had been working at Xerox in difficult times when the only option for the company was to learn from the best (their Japanese colleagues at Fuji-Xerox) and improve their practices.

However, those were also the days of the dawn of the Office of Water Services (OFWAT). Regulation was introduced in England and Wales in 1989 following the privatization of the water sector. Yardstick competition (the use of metrics to compare the performance of the different water companies) was soon chosen as a key regulatory tool. Such artificial competition was to compensate for the lack of a real market tension between water companies, and therefore drive them to improve performance.

It is hardly surprising, that when the AWWA Research Foundation issued a report on benchmarking (Kingdom and Knapp, 1996) the universe described for water utilities was ruled by Xerox benchmarking and the yardstick competition created by OFWAT. However, the language used to describe that universe proved to be far more controversial than the tool itself.

The AwwaRF report, well written and easily understandable, coined the terms *metric* and *process* to qualify what had simply been *benchmarking* until that date. However, while the text had an overwhelming success in the dissemination of these terms, the underlying concepts did not travel as far.

Metric benchmarking made reference to the comparison of key performance indicators. This was (and still is) the basis for the system used by OFWAT and by most of the water industry regulators in the world. However, metric benchmarking soon became appealing to a completely different type of user: utilities who wished to determine their competitive level by screening their performance areas for strengths and weaknesses and comparing them to the performance of other utilities.

Process was used to identify "Xerox benchmarking" as described by Camp. This method identifies an organization that is best in class in a particular process or business area; and after determining which factors are key to the success, adapts the best practices to improve performance.

The two techniques shared some similarities. They both needed the participation of several utilities. They both involved some sort of comparison and, above all, metrics were a crucial first step for Xerox benchmarking (although they should never become the main aim of the project).

However, this picture that clearly separated the concepts and techniques (as presented by Kingdom and Knapp) never made it to the general public. Both concepts have often been confused since then, and the actual words (especially "process") have led to a good number of misunderstandings as people tried to interpret what the terms implied.

And so, for a whole decade, we witnessed the appearance of dozens of papers and technical references in which the terms "metric benchmarking" and "process benchmarking" were used with different meanings, representing all shades of grey. This lack of consistency in the definitions was even present in technical working groups and international projects, and their name no longer helped to assess the nature of an effort without a thorough explanation of the work. It could be argued that no one was wrong in this story and that words are just words. But the truth is that benchmarking specialists need a common language to be able to forget about words and start working on the techniques.

1.3 A NEW BENCHMARKING FRAMEWORK

One of the main objectives of this manual from the very beginning was to clarify the difference between metric benchmarking and process benchmarking, an issue that the industry needs to leave behind. Endless hours have been devoted in the past decade to discuss whether a project was a metric, a process or even a benchmarking project at all.

The bottom line to all those discussions is that the value of a certain initiative or project does not really lie in its name, and whether it can be called a benchmarking project. Once again it must be stressed that benchmarking is only one of the many available tools, and results are much more important than the tool itself.

Unfortunately, even we soon found out that the line between metric and process is just too difficult to draw. At the time of writing this manual, too many years had spanned with many uses of both terms to allow reconstruction of their meaning. A similar situation was faced by the IWA Water Loss taskforce a decade ago (Lambert and Hirner, 2000) when they decided to discontinue the use of the term *unaccounted for water*[3]. That was a successful initiative, and despite the fact that the term is still used today, the terminology proposed by both IWA and AWWA regarding water loss has become a common language shared by the water industry worldwide.

With that model in mind, we decided that this manual should be the ideal vehicle to present a new benchmarking framework.

In the following pages, you will be provided with a complete picture of benchmarking in the water industry in the world without the use of the terms *metric* and *process*.

As a matter of fact, let us clearly stress that point in the following sentence:

> **The IWA Specialist Group on Benchmarking strongly recommends abandoning the use of the terms "metric benchmarking" and "process benchmarking". Instead "performance assessment" and "performance improvement" should be considered consecutive components of benchmarking.**

To better illustrate the concepts of performance assessment and improvement, Figure 1.1 maps most of the practices qualified as benchmarking in the water industry. Any current benchmarking project can be placed in the figure according to the objectives pursued, the techniques used and also taking into

---

[3] The term *unaccounted for water* was kept out of the IWA water balance (now adopted by AWWA as well) due to the confusion it brought to the industry. Once again, the meaning of the words composing the term prevented any clarification from a standard definition. In the end, and despite the popularity of the term, it was decided that its absence was less harmful than the confusion it brought.

account the level of detail (in other words, the power of the magnifying glass or how deeply in detail is the utility being studied).

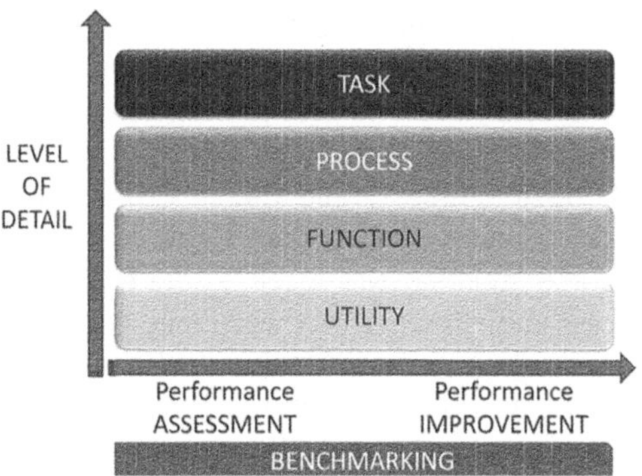

Figure 1.1 The performance assessment and improvement model

This new framework can also accomodate the regulatory activities in the industry, which are usually focused on comparative performance assessment, but strictly speaking do not constitute benchmarking. Therefore, it would be advisable to use terms like yardstick competition or comparative performance assessment for those regulatory efforts[4].

As a matter of fact, one of the main problems encountered to date when defining benchmarking was that different projects focused on a different level of detail within the utility (vertical axis). While some of the projects studied the utility as a whole (lower level in the figure) other projects went into greater detail and focused on the main functions that conform the core business of water utilities, the processes that compose those functions or even the elementary tasks in which those processes can be broken down[5].

---

[4]The systems used by regulators are not focused on improving the performance of utilities and usually stop short of the performance improvement step. This should not be taken to mean that regulators do not encourage the adoption of best practices by the regulated utilities. On the contrary, they often do. However the performance improvement actions are not part of the regulatory framework.

[5]Organizational levels receive different names depending on the reference. The IWA manuals on performance indicators make reference to main functions, partial functions and sub-functions.

Good examples of these different approaches can be found in current projects of both performance assessment and performance improvement. For instance, most regulators (like OFWAT in England and Wales or ERSAR in Portugal) assess water utilities at the utility level, and although there may be a specific focus in some key strategic areas, the assessment is rarely done at the function or process level. On the other hand, it is quite common for industry driven projects to focus on functions (customer service, asset management, etc.) and to go down to the process and task level in order to find answers to the "how to improve" question, for instance the benchmarking projects promoted in Australia by the suppliers' association, WSAA.

Example

In this WSAA example, a thorough examination was conducted for participating utilities. The "Customer Service" function was evaluated, including the Call Center, Field Services, Credit and Collection, Billing and Accounting and Back Office "Processes".

Activities included the practices performed in carrying out this process. Cost and service level performance data was assembled and assessments were conducted to determine the relationship between the work practices used, costs, and service levels achieved.

	Cost and workload analysis	Productivity drivers	Service level measures	Practices
Call centre	• Total call centre cost per customer • Total call centre cost per equivalent task • Equivalent task per customer	• Equivalent task per FTE	• Average speed of answer by CSR • Average speed of answer by IVR • Percentage of calls answered in 30 sec/60 sec. • Percent of calls resolved on first contact	• Internal performance standards/ goals • Customer transaction fulfilment activities

(Continued)

Example (*Continued*)

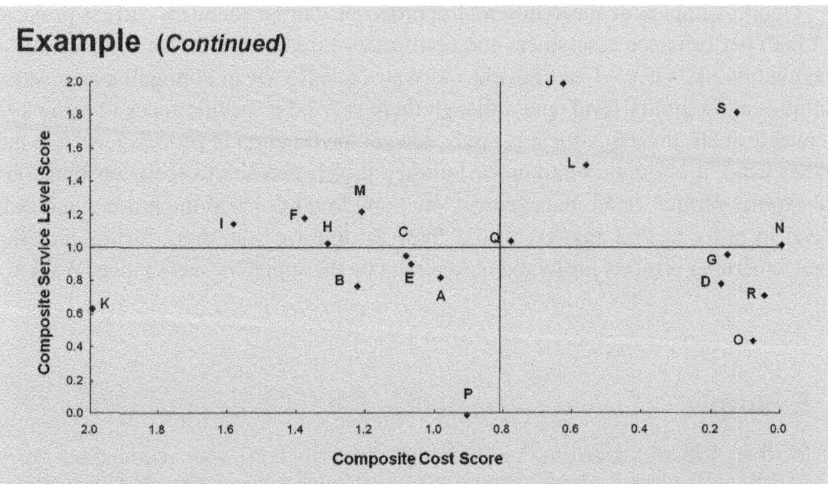

In addition, data was normalized and comparisons were made between utilities. The graphic above displays each utility's performance, based on cost and level of service information. The relative comparisons are evident. Utilities located in the upper right quadrant are those with the lowest costs and the highest levels of customer service. An examination of the practices producing these results was carried out and communicated in a best practice workshop to better position other utilities to improve their performance. Also, each utility was able to identify desired utility improvement targets and stretch goals to achieve cost and level of service goals.

The assessment and improvement of performance at different levels of detail within a utility is perfectly logical depending on the needs and objectives of each project, and numerous successful examples can be found in the literature.

In each one of the levels of detail described in Figure 1.1 is possible to carry out a performance assessment. For instance, a funding institution will try to assess the overall utility performance or at least its overall financial or service performance. However, a plant engineer may be quite interested in both the assessment and the improvement of the performance for a single task within a very specific process (membrane cleaning in a microfiltration plant).

Additionally, a first and necessary step in benchmarking is the assessment of how efficiently or to which standard a certain utility operates, or a function, task or process is carried out. Such assessment is usually done by means of performance indicators which need to be compared to a reference to obtain the assessment[6] (the reference can be a fixed one – e.g. a standard or target – or the performance of a peer – e.g. another water utility –). We call this the comparative performance assessment phase.

Once performance has been assessed, a logical follow through is to undertake actions to improve it. And hence, by means of identifying and adapting the best practices of those who are better (or the best) at a certain function or process, it is possible to improve performance. We call this the performance improvement phase, and it requires the participation of several utilities or benchmarking partners (even from other industries) in order to gather additional information that will lead to the identification and adaptation of best practices.

These two clearly differentiated phases (performance assessment and performance improvement) are the necessary parts of benchmarking. And therefore, benchmarking can be defined as "a tool for performance improvement through systematic search and adaptation of leading practices".

This new framework can easily accommodate any project carried up to date within the water industry. As a matter of fact, a very interesting exercise consists in identifying the scope of existing projects in Figure 1.1.

For instance, in Figure 1.1, metric benchmarking projects (as defined by Kingdom and Knapp in 1996) would occupy the left half of the figure, while process benchmarking would be represented by the right half (Figure 1.2).

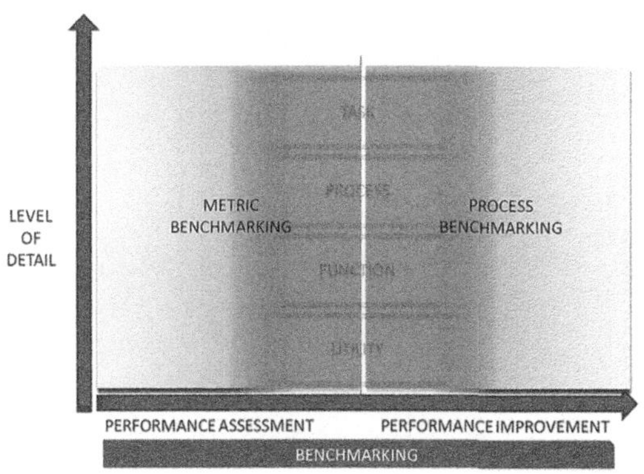

Figure 1.2 Metric and process benchmarking as described by Kingdom and Knapp (1996)

---

[6] However, the sole use of performance indicators and the consequent analysis would not constitute benchmarking.

This division, as seen in the figure, was made irrespectively of the level of detail of the project, and therefore a vertical axis would not be necessary in order to define the technique used. In the case of the definitions created by Kingdom and Knapp, metric benchmarking would fall completely under what we call now the performance assessment phase, while process benchmarking would belong to the performance improvement stage.

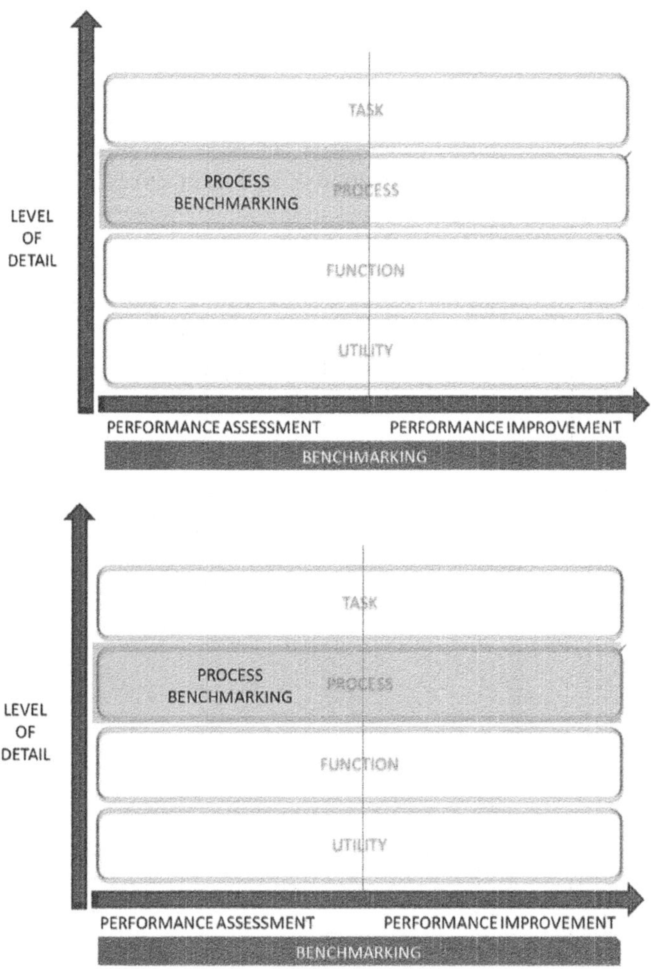

Figure 1.3 Process benchmarking as understood by several projects worldwide

Some of the confusion in the past originated in the understanding of the term "process benchmarking". The comparison of performance at process level was sometimes referred to as "process benchmarking" (Figure 1.3). This understanding of the definition was not an isolated case. However, in some of these projects, the objectives did not include performance improvement (a great difference from the original definition). The comparison of Figure 1.2 and Figure 1.3 clearly shows that those two concepts of "process benchmarking" were quite far from each other and certainly needed clarification.

Once projects are represented in this scheme, the origins of misleading terms can be easily traced and their true nature understood. For instance, most of the former metric benchmarking projects (which collected performance indicators and compared their values) were focused on performance assessment at the utility level. They would occupy the left side of the figure (performance assessment) and only the lowest bar (utility).

It should be made clear that the value of a project does not depend on whether it is a benchmarking project or not. Most of the projects to date have been valuable to participating utilities, pointing out deficiencies in their data handling procedures and unveiling weaknesses or potential improvement areas. The aim of this new terminology is not to differentiate between good and bad techniques but to provide clear definitions that will allow industry members to communicate more effectively. Both performance assessment and performance improvement are valuable tools that any utility should consider to become more efficient. And in any case, benchmarking is only one of the different tools that can be used to achieve utility improvement.

Performance assessment	Benchmarking (performance assessment & improvement)
• Encourages collecting the right information and improving data quality • Gives a first insight on where improvement may be found • Is one of the main tools for regulation worldwide • Many performance assessment projects later evolve into performance improvement efforts	• Includes more than just the comparison of metric figures • Aims to compare equivalent functions and practices to establish relative position. • Contains both quantitative, as well as strong qualitative components • Facilitates and involves improvement action planning from comparisons and exchange of experiences with peers (workshops, utility visits etc.)

1.4 WHY SHOULD YOU BENCHMARK?

Finding the right motivation to benchmark is an unavoidable first step in any project. A benchmarking project is a demanding endeavour, which needs to involve all levels of any organization and receive full support from them (including senior management). Finding the answer to the title question is the unavoidable first step that any organization should tackle in benchmarking.

In the past 20 years, many benchmarking projects have been undertaken in the water industry all over the world. In many occasions, the drive to benchmark was initiated outside the utility, and benchmarking was seen as a logical option after one of the following happened:

- Public debate about liberalisation or privatisation of public services.
- Demand for more transparent and efficient public services.
- Political pressure for full cost recovery.
- Requirements for large investments to improve the service in terms of coverage and quality, which in their turn require smart innovations by the water industry to keep charges at a reasonable level.

However external drivers are usually not sufficient to initiate benchmarking (excluding regulatory obligations). Usually, the utility management, with its operational responsibility for the service, has its own drivers to benchmark. The natural need to continuously improve the organisation and its products can be facilitated in benchmarking projects that provide a detailed insight in the performance and identify areas and means of improvement. The relative position of a utility against its peers can help determining the sense of urgency for taking measures in the analysed areas.

However, if something has also been made clear during all these years is that many utilities are reluctant to enter the benchmarking arena. Therefore, we should also acknowledge that there may be reasons for a utility to refrain from joining a benchmarking program:

- The utility is considered to be unique and not comparable.
- The company is being restructured or merged, so there is no stable situation for assessing performance.
- There is a lack of reliable data to submit.
- There are not enough available resources (budget, manpower).
- There are doubts on the added value of the program.
- The suggested methodology is too complicated.
- There is no guarantee for confidentiality of individual performance data.

Therefore, these issues shall be kept in mind when a benchmarking activity is prepared.

The benchmarking scene is not limited to utilities. As a matter of fact, the industry itself has sometimes been less interested in benchmarking than other stakeholders:

1.4.1 Governments/regulators

Governments are politically responsible for water services. Regulators, when present, focus more at the individual utility level. In any case, comparative performance assessments can help both governments and regulators to introduce artificial competition in a sector which constitutes a natural monopoly and to put pressure on utilities to raise efficiency and transparency.

Governments & regulators – Why assess performance?

- Political responsibility for the regulation of the water sector.
- Need to guarantee:
  - Appropriate levels of service and compliance with applicable standards.
  - Sustainable and efficient operations, transparency of the water industry.
  - Appropriate organisation/regulation of the water market (natural monopoly).
- Benefits from comparative performance assessments:
  - Insight in performance of the water sector (compliance, service levels, investments etc.).
  - Pressure on utilities for efficiency.

1.4.2 Customers

Customers and consumer organizations are usually focused on obtaining a good service and a good product while paying a fair price. Additionally, questions regarding affordability for those with the lower incomes are often a main concern.

Benchmarking is an excellent tool for consumers, and so is comparative performance assessment (it is not difficult to find consumer driven reports comparing several utilities).

<table>
<tr><td rowspan="2">Customers – Why benchmark?</td><td>

- Require adequate water services with appropriate level of service.
- Expect affordable prices and value for money.
- Benefits from comparative performance assessments:
  - Insight in the performance of the local utility.
  - Service provided by a utility continuously seeking efficiency and in competition with peers.
  - Comparative assessment of value for money.

</td></tr>
</table>

1.4.3 Owners/shareholders

The owners and shareholders are legally responsible for the utility. As a result of such responsibility, and to comply with applicable regulations, they need insight in the utility's performance, its efficiency and the magnitude of the financial or other risks undertaken by the utility.

Benchmarking and comparative performance assessment are excellent tools to assess those magnitudes and place them in perspective when compared with others. Additionally, benchmarking demonstrates that there is a culture within the organization to continuously improve and become more efficient.

<table>
<tr><td rowspan="2">Owners/Shareholders – Why benchmark?</td><td>

- Legally responsible for the utility.
- Require insight in
  - utility performance
  - efficiency in the operations
  - financial sustainability
  - proper risk assessment.
- Benefits from benchmarking:
  - Insight in performance of own utility compared to peers.
  - Continuous improvement in the utility.

</td></tr>
</table>

Chapter 2

Performance assessment basics

Assessing performance is a natural need in any human activity. We are often trying to determine how we compare to other people and whether we could be better. As a matter of fact, in order to assess performance we usually tend to compare to others or to ourselves over time. For this same reason, it is difficult to determine if someone is being good at something if there are no previous references (whenever a new sport is created, world records are often improved and it is difficult to know how good a certain performance is). However, this approach becomes problematic when two situations are difficult to compare side by side.

Performance assessment of water utilities is not an easy job. The amount of data present in a single utility can be overwhelming. As a matter of fact, the Geographic Information System (GIS) of a utility can hold terabytes of data, and yet still not capture all the details of the system. The conclusion is that reality is very complex and all our efforts to reproduce it require some sort of simplification. In this simplification process we find two contradictory needs: on one hand, a higher level of detail delivers a more faithful representation of reality. Computers allow us to construct models of reality which get more and more complex everyday, such as the GIS mentioned before. On the other hand, large quantities of data are not always the best option to make decisions and do not constitute information (we could define information as data which enable us to make decisions). This is why CEOs often require short reports instead of thousands of pages and we feel more comfortable using summary lists than using large quantities of raw data for decision making.

For those same reasons, performance assessment could be described as the art of simplification: the more condensed the data is, the better; but an excessive simplification of the picture may not provide sufficient information to make sound decisions.

Indicators are a great tool to assess performance. The traditional ratio combines at least two relevant variables measured in the real world and provides significant information. By combining the adequate indicators a general picture of reality can be achieved. An indicator is a very intuitive tool, and is easily understandable. Additionally, indicators facilitate comparisons, as denominators usually provide a size or quantity reference.

Despite the fact that indicators are accessible and easy to understand, creating a good indicator is not always an easy task. A good indicator needs to fulfil certain characteristics, as described in the IWA manuals of best practice on performance indicators for water supply (Alegre *et al.*, 2006) and for waste water (Matos *et al.*, 2003):

"Individually, a performance indicator (PI) should comply with the following requirements:
- Be clearly defined, with a concise meaning.
- Be reasonably achievable (which mainly depends from the related variables)
- Be auditable.
- Be as universal as possible and provide a measure independent from the particular conditions of the utility.
- Be simple and easy to understand; and
- Be quantifiable so as to provide an objective measurement of the service, avoiding any personal or subjective appraisal.

Collectively, PI should comply with the following requirements:
- Every PI should provide information significantly different from the other PI in the system.
- Definitions of the performance indicators should be univocal (this requirement is made extensive to its variables)
- Only such PI should be established which are deemed essential for effective performance evaluation."

Performance indicators are only useful when compared to a reference. The value of an indicator, without any additional information attached may not be meaningful. Therefore, when designing a performance assessment system, the comparison method must be clear. Indicators can be used for:
- **Assessing the fulfilment of objectives/targets**. Strategic objectives need to be controlled to assess if they were achieved. Indicators are a great tool to measure whether a certain target has been met and to which extent, and therefore if the strategy used to meet them was the right one.
- **Trend analysis**. Whenever the focus is on the evolution with time of the utility or parts of it, indicators allow providing trends and even predictions.

In this case, the indicators are being compared to previous values of the same indicators obtained in the past, and obviously deliver information about the evolution of performance and whether it is improving or not. A deeper analysis may try to take into account the variation of several indicators and system variables to explain the changes in performance in the system.

- **Peer comparison**. A natural follow-up to any indicator system is to try to compare the indicator values with those obtained by another utility. As a matter of fact, most utility managers using indicators for internal purposes have tried at one stage or the other to search for published reference values from other utilities. In order to assess whether a utility is efficient and if its performance is better or worse compared to other utilities, indicators can be used. In this case, the required system is more complex, as the analysis of the results needs to take into account size and context differences.

Regardless of the purpose, the performance assessment system needs to be well designed and tailor made for its objective. The IWA manuals on performance indicators for water supply (2nd edition) and waste water provide a structure that may prove to be a valuable guide when building up such a system.

2.1 THE IWA PERFORMANCE INDICATOR SYSTEM

Since its publication in 2000 (and its further revisions in 2003 – wastewater–, and 2006) the IWA system of performance indicators has become the industry standard on the topic. The manuals provide a long list of indicators (over 150) and an even longer list of variables which are needed to calculate the indicators. These indicators provide a useful shopping list that can be used by utility managers when designing their performance assessment system.

Once a certain area or process is identified as needing to be measured in a utility (e.g. real losses) the IWA manual provides with several possible indicators which can be used to that purpose. There are several advantages to using one of the PI proposed by IWA:

- All indicators fulfil the requirements previously listed in this chapter. Additionally, the indicators published by IWA went through a revision process with hundreds of contributions guaranteeing they are valuable measures of performance.

- The indicators and the corresponding variables are well defined. Often, performance assessment projects find out that the definitions used for their indicators at the start of the project are not solid enough to guarantee that all users understand them in the same way. IWA definitions may not be perfect, but they are detailed enough to guarantee that the debate will only concern project specific details.

- The proposed indicators have become an industry standard. This means that there are good chances that, once an indicator is chosen, someone else in the world will be using the exact same indicator. This increases the probability of both obtaining reference values from others and/or finding other utilities to which compare performance. This same approach has been taken by some of the main benchmarking projects in the world like the European Benchmarking Network. Therefore, for any potential participant, it would be easier to join such a project if the data collection procedures and the indicator definitions were already in use in the utility. In addition, in any benchmarking project it is extremely important to use precise definitions which are shared by all participants. The IWA set of indicators will allow either choosing those indicators off the shelf, or at least using them as a base for modification to create new ones.

However, as important as all those definitions of indicators and variables are, even more important is the structure of a performance assessment system provided by IWA. The framework allows adding, replacing or modifying indicators with the assurance of the system being coherent and compatible with other systems in the world.

In other words, IWA acknowledges the fact that the list of over 150 indicators cannot provide an answer to all questions. Users are expected to need indicators for additional purposes, and in such case, they should create them from scratch. However, the IWA framework provides guidelines on how to structure a performance assessment system, so the new indicators can integrate seamlessly with those selected from the IWA list.

A system of performance indicators is comprised of a set of performance indicators and related data elements which represent real instances of the utility (Figure 2.1). The classification of these data elements depends on the active role they play. These are the main elements of the IWA performance indicators system.

2.1.1 Data elements

A data element is a basic datum from the system which can either be measured from the field or is easily obtainable. Depending on their nature and role within the system, data elements can be considered variables, context information or simply explanatory factors.

2.1.1.1 Variables

A variable is a data element from the system that can be combined into processing rules in order to define the performance indicators. The complete variable consists of a value (resulting from a measurement or a record) expressed in a specific unit, and a confidence grade which indicates the quality of the data represented by the variable.

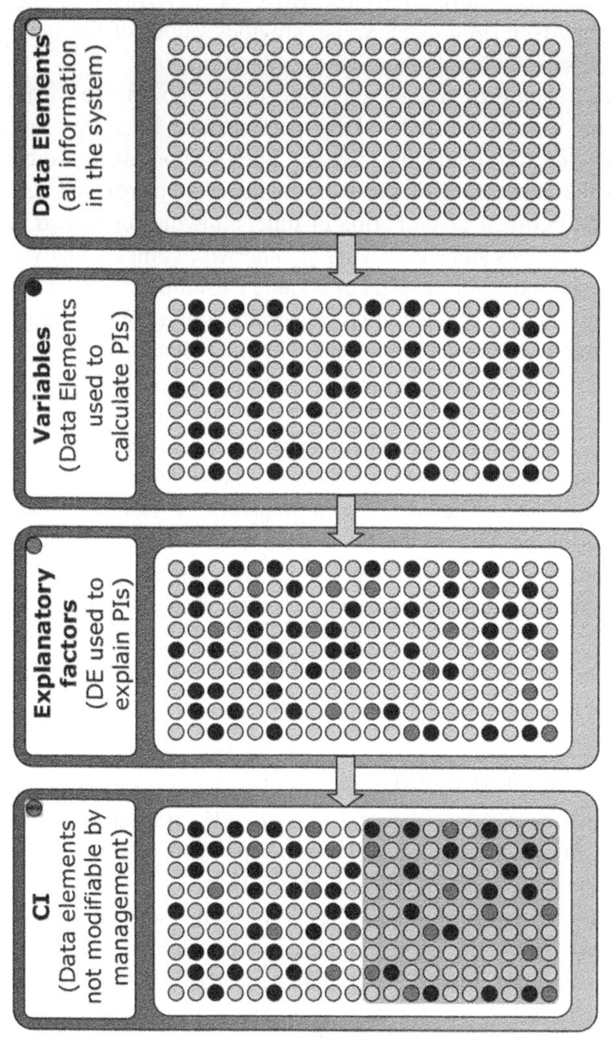

Figure 2.1 Components of the IWA Performance Indicators Systems

2.1.1.2 Performance indicators (PI)

Performance indicators are measures of the efficiency and effectiveness of the delivery of the services by an undertaking that result from the combination of several variables. The information provided by a performance indicator is the result of a comparison (to a target value, previous values of the same indicator, or values of the same indicator from other undertakings).

Individual PI should be unique and collectively appropriate for representing all the relevant aspects of undertaking performance in a true and unbiased way, thus reflecting the managing activity. Each performance indicator should contribute to the expression of the level of actual performance achieved in a certain area and during a given period of time, allowing for a clear comparison with targeted objectives and simplifying an otherwise complex analysis.

A performance indicator consists of a value (resulting from the evaluation of the "processing rule") expressed in specific units, and a confidence grade which indicates the quality of the data represented by the indicator.

Performance Indicators are typically expressed as ratios between variables; these may be commensurate (e.g. %) or non-commensurate (e.g. $/m^3). In the latter case, the denominator shall represent one dimension of the system (e.g. number of service connections; total mains length; annual costs), to allow for comparisons. The use as denominators of variables that may vary substantially from one year to the other, particularly if not under the control of the undertaking, shall be avoided (e.g. annual consumption, that may be affected by weather or other external reasons), unless the numerator varies in the same proportion.

A clear processing rule shall be defined for each indicator, specifying all the variables required and their algebraic combination.

2.1.1.3 Context information

Context information are data elements that provide information on the inherent characteristics of an undertaking and account for differences between systems. There are two possible types of context information:

- Information describing pure context and external factors to the management of the system. These data elements remain relatively constant through time (demographics, geographics, etc.) and in any case are not affected by management decisions.
- Some data elements on the other hand are not modifiable by management decisions on the short and medium term, but the management policies can influence them on the long run (for instance the state of the infrastructures of the utility).

Context information is especially useful when comparing indicators from different utilities.

2.1.1.4 Explanatory factors

An explanatory factor is any element of the system of performance indicators that can be used to explain PI values, i.e., the level of performance at the analysis stage. This includes PI, variables, context information and other data elements not playing an active role before the analysis stage.

The use of performance indicators should always be linked to the establishment of a proper performance assessment system, in which all the above mentioned elements are present and defined, and aimed to fulfil a clear objective or obtain information on specific areas or issues.

A list of all the indicators, variables and context information items in the IWA proposals is included in Annex C.

2.1.1.5 Confidence grading

One last important issue regarding the definition of a performance assessment system is the quality of data. In most cases, quality of the data used to feed the indicator system is not even recorded. Quality is either taken for granted or not considered very important. However, if the main use of a performance assessment system is decision making it is difficult to imagine data quality not being relevant. For instance, making a crucial decision on an indicator with a value of $20 \pm 1\%$ in error, is completely different from having to make that same decision if the result is $20 \pm 100\%$.

The IWA system measures data quality in terms of accuracy and reliability. The reliability of the source accounts for uncertainties in how reliable the source of the data may be, i.e., the extent to which data source yields consistent, stable, and uniform results over repeated observations or measurements under the same conditions each time.

The accuracy accounts for measurement errors in the acquisition of input data, i.e., the closeness of observations, computations or estimates to the true value as accepted as being true. Accuracy relates to the exactness of the result, and is distinguished from precision which relates to the exactness of the operation by which the result was obtained.

Practice shows that, in general, data providers do not have detailed information on reliability and accuracy, but are able to provide informed guesses, if broad bands are adopted (Table 2.1).

Table 2.1 Recommended accuracy bands

Accuracy band	Associated uncertainty
0–5%	Better than or equal to +/−5%
5–20%	Worse than ±5%, but better than or equal to +/−20%
20–50%	Worse than ±20%, but better than or equal to +/−50%
>50	Worse than ±50%

Recommended bands for the reliability of the source are:

Table 2.2 Recommended data source reliability bands

Reliability band	Definition
***	Highly reliable data source: data based on sound records, procedures, investigations or analyses that are properly documented and recognised as the best available assessment methods.
**	Fairly reliable data source: worse than ***, but better than *.
*	Unreliable data source: data based on extrapolation from limited reliable samples or on informed guesses.

2.2 IMPLEMENTATION STEPS

Overall, setting up a proper performance assessment system based on indicators is not as easy a task as most people may think. Experience shows that a systematic approach is necessary to create a balanced group of indicators that can be used for a certain purpose. One of the main ideas to bear in mind is that measuring is not an objective in its own, but just the start of a process to achieve better performance. It is important to lay good foundations in any performance assessment project, and that means to clearly identify the project objectives before a single indicator is even mentioned.

This does not mean that indicator selection is not important. On the contrary, performance assessment results greatly depend on which indicators are selected. After all, we are trying to represent reality through a simplified model (the indicator system). However, oversimplifying the model or simply focusing on the wrong parts of the utility, may lead to an incorrect interpretation of reality.

The selection of indicators needs to be aligned with strategic objectives and with the strategies which are devised to reach those objectives. In the case of projects with several utilities, the indicators should be chosen according to the common objectives of all participating utilities. This is usually a good starting point. Once the objectives are clearly identified, the selection of indicators becomes a much easier task.

Figure 2.2 shows the implementation steps proposed for a performance assessment system by IWA.

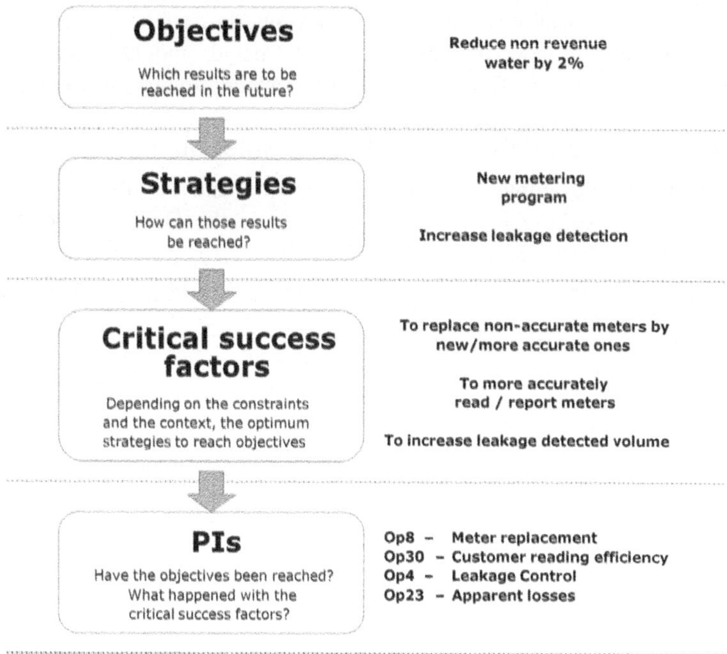

Figure 2.2 PI as a part of a performance measurement system

2.3 INDICATORS SELECTION

As it has been mentioned before, the last step of implementation (choosing the indicators) is a crucial one. Even when we are sure of what we want to measure, we are not certain of finding the right measurement tool.

The construction or selection of a performance indicator often implies some bias. This is often discovered in projects where different participants have strong preferences over one indicator or the other. Once again, the model chosen to represent reality may favour one perspective over the other, and one simplified version of the world may provide better results than another.

The IWA indicator definitions were created trying to avoid these bias factors. Therefore, the selection of one of the proposed indicators will always constitute a good starting point in indicator selection. However, an IWA indicator only guarantees a good measure and it must be studied for each case whether it is perfect for a purpose.

Besides using indicators which have been properly designed, it is important to take into account how indicators are constructed, and how the different variables composing an indicator may affect its capacity to represent reality.

Example – Choosing the right indicator: water quality

A good practical example on indicator selection can be illustrated by the European Benchmarking Co-operation (EBC) and the Dutch Association of Water Companies (VEWIN).

The EBC chose one of the IWA quality indicators (QS 18 "Quality of supplied water", Figure 2.3):

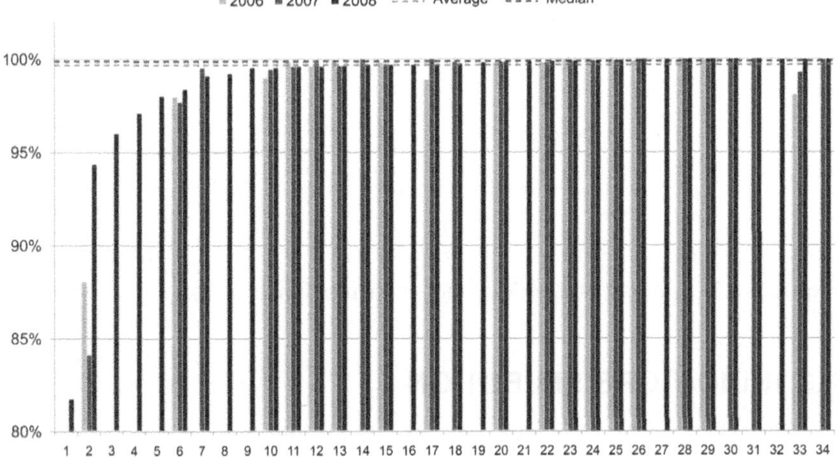

Figure 2.3 Quality of supplied water for 25 European water utilities in 2007. Source: EBC, 2008

The indicator shows the percentage of drinking water tests that comply with the applicable standards. The fact is obviously relevant to all consumers and the indicator itself has never been contested. However, in the EBC context, most utilities comply with all quality standards all the time. As a result, the comparison of indicator values provides little information.

As an alternative, the "Water Quality Index" (WQI) was developed by the Dutch Association of Water Companies (Vewin) for its national benchmarking program (Figure 2.4). The metric takes values between 0 (no substances at all in the water) and 1 (drinking water according to legal standards). In simplified terms, the lower the value, the better the water.

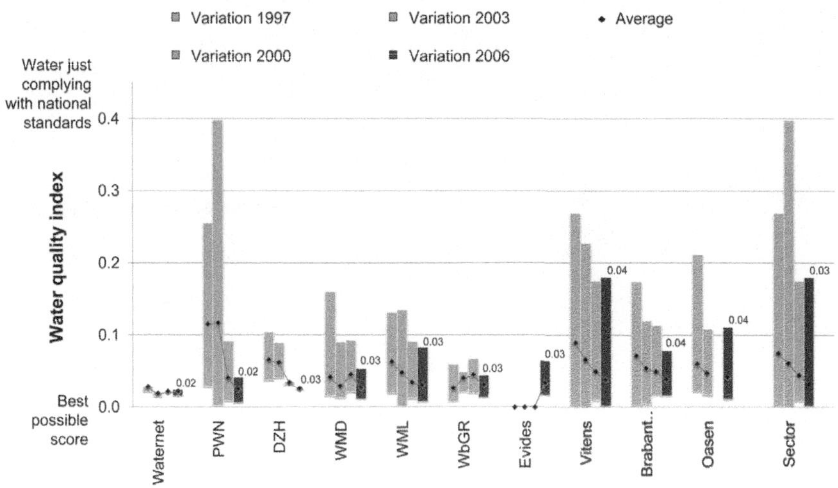

Figure 2.4 Water Quality Index 2006 for Dutch water companies. Source: Vewin, 2007

The WQI would not be considered an indicator as formally defined by IWA, and is rather an aggregated index. 33 key water quality parameters are considered in its calculation.

However, in a specific environment like the one defined by regulations in The Netherlands, the WQI provides more useful information than IWA's QS18 indicator. On the other hand, aggregated indexes are more difficult to define as the scope broadens (in this particular case, the relevant parameters and their relative importance may change depending on the local circumstances).

The lesson to be learnt from this example is that indicators should be aligned with strategic objectives to provide relevant information. In the case of QS18, it is a useful measure when the actual compliance with water quality regulations should be considered. However, all utilities in The Netherlands have a 100% compliance with regulations most of the time, rendering QS18 quite useless. In this case, the focus turns to the actual quality of water (not the regulations) and that is the aim of the WQI. For instance, two utilities may have a 100% compliance with legal standards and yet the quality of the water supplied by the two utilities can still differ a lot.

Chapter 3
The IWA benchmarking process

3.1 THE BENCHMARKING PROCESS

The first mainstream reference to benchmarking was probably the 1989 book "Benchmarking – The search for Industry Best Practises that Lead to Superior Performance" by Robert C. Camp. His description of the modern concept of benchmarking is based on the case of Xerox Corporation, where he used to work. In the 1970's, this US copier manufacturer heavily lost market share to Japanese manufacturers. For Xerox Corporation this came as quite a surprise, for the company was increasing its productivity and did not pay much attention to developments outside the organization.

Looking for explanations why sales had decreased, Xerox started a comparative analysis of copiers from different competitors in 1979. Functional specifications were compared, copiers were dismantled and mechanical parts investigated. More extensive benchmarking followed by comparing copiers of other Japanese manufactures, including its Japanese subsidiary Fuji-Xerox. The results confirmed significantly higher production costs in the US. As it turned out, Japanese companies – including Fuji – sold copiers at Xerox's production cost. This marks the birth of the modern concept of benchmarking, which in 1981 was introduced at corporate level in Xerox.

Although benchmarking was developed as a tool for business improvement to gain back market share in a competitive environment, it can also be applied in the water industry which generally lacks competition. After all, the concepts applied in benchmarking are universal and their application does not depend on the type of industry or service.

It must be emphasized that benchmarking is not the only tool for improving water services. Other options include process optimization, business process redesign, restructuring, merging utilities, etc. However, over the past decade many cases have proven benchmarking to be a powerful management instrument to achieve improvements in the water industry.

3.1.1 The benchmarking concept

As introduced in chapter 1, benchmarking consists of two consecutive components. The first step, performance assessment, aims at analyzing performance, comparing it with other organizations within or outside the industry, and identifying performance gaps. The next step, performance improvement, is designed to find improvements by learning from the leading practices and adapting them to the own situation.

Benchmarking usually is organised in projects ("exercises") with start and finish dates. However, from a management point of view, benchmarking should not be considered a single, isolated action but a continuous process. The search for better practices never ends. Legal requirements, customer demands, water technology, ICT and asset management techniques (for instance) develop rapidly and utility management cannot afford to lean back. What was considered a good practice yesterday, could well be outdated today. Therefore, utilities need to permanently search for opportunities for further improvement to assure that their customers get the best value for money and shareholders do too. Therefore, benchmarking should follow the plan-do-check-act principle (Figure 3.1).

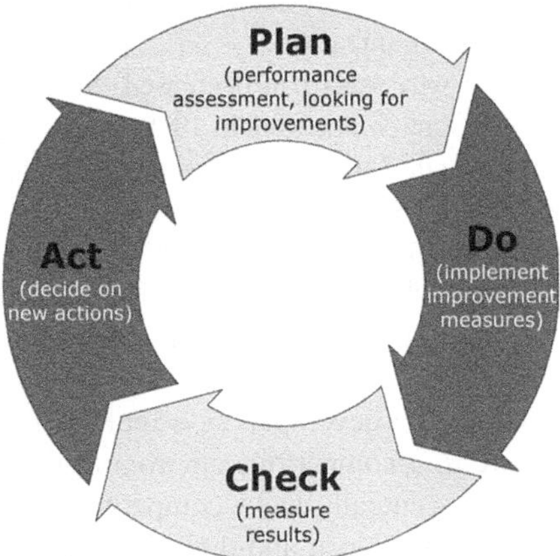

Figure 3.1 Plan-do-check-act cycle

Benchmarking should be embedded in the (annual) business planning cycle. This way, it can be connected to the strategic objectives of a utility, avoiding to turn it into a stand-alone project.

3.1.2 The IWA benchmarking process

Although almost every benchmarking reference in the literature has its own benchmarking process (with different number of steps) they all follow the same principles. A typical benchmarking process with six different steps is presented here (Figure 3.2).

1. Project planning.
2. Orientation, training and project control.
3. Data acquisition & validation.
4. Data analysis & assessment reporting.
5. Improvement actions
6. Review of improvement actions.

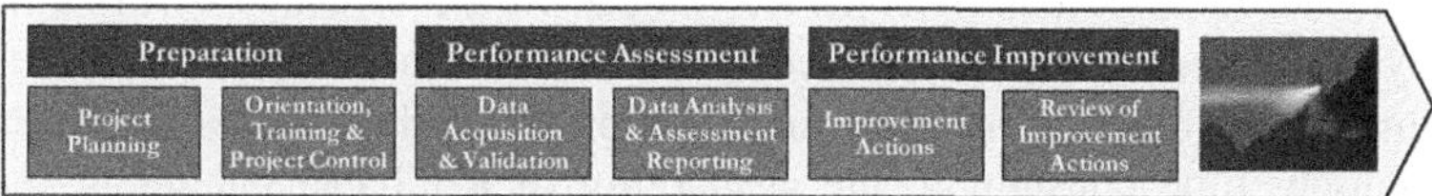

Figure 3.2 The IWA benchmarking process

These steps are briefly described here, and they will be discussed in more detail in the following chapters.

3.1.2.1 Project planning

At the start of a benchmarking project, the scope and level of detail are determined based on the demands and needs of the interested utilities. The performance assessment model and the data requirements also need to be elaborated to show participants what they can expect, and to estimate project resources. Based on this information, a detailed project plan with budget and planning can be drafted.

In case of a closed group a "go/no go" decision can be made based on the project planning. In case of an open project, interested utilities are invited to participate at this stage and based on their response the project may or may not be launched.

Project planning is further covered in chapter 5.

3.1.2.2 Orientation, training and project control

Before starting to benchmark, all staff involved in the project needs to be prepared. The objectives of the exercise and the project plan should be clear. Furthermore, staff needs to be informed about the methodology and data requirements and trained in the data collection methods that will be applied in the project.

These considerations include the staff from participating utilities and staff from the project team (organizing body and/or commissioned third parties).

For further details, see chapter 6.

3.1.2.3 Data acquisition and validation

One of the most time consuming activities in a benchmarking project is the data acquisition by the participants. This step requires significant efforts from the participants, depending on their experience, availability of the information and accessibility. The role of the project team in this step is to assist utilities in clearing up methodology issues and definition problems and to secure meeting deadlines.

When the required data is collected, it needs to be validated by the utilities and by the project team, for instance by looking at consistency with data from previous years, outliers, on-site visits or auditing. Although this activity may be intensely time consuming, the availability of a reliable dataset is key to successful benchmarking. Participants in a benchmarking project expect good quality comparisons and, accordingly, proper identification of performance gaps as these are the triggers for improvement actions.

Data acquisition and validation issues can be found in chapter 7.

3.1.2.4 Data analysis and assessment reporting

Once data are validated, they are analyzed, performance indicators are calculated and performance comparisons are made between the participants. In this stage possible remaining errors in the dataset can be identified and cleared up to improve data quality. Performance gaps are then determined and explained (if possible) keeping in mind the differences in the operating environment of the utilities.

The result of this step is a draft report (at individual and/or group level) with the preliminary results of the performance assessment. This is the basis for discussing performance differences with the participating utilities in a workshop.

After discussing the preliminary results of the performance assessment in a workshop, possible errors and further explanations on performance gaps or differences in the operating environment of utilities are processed. Final reports on the performance assessment are produced and distributed to disseminate the results within the company and to its stakeholders. These assessment reports can be supplemented by improvement action plans after the upcoming steps.

For further details on all reports, see chapter 8.

3.1.2.5 Identification, prioritization and implementation of improvement actions

One of the most important activities for reaching the final goal of a benchmarking exercise involves taking theory into action. Based on the performance assessment and the knowledge that is available in the network, the project team and the participating utilities jointly try to discover best practices, present and discuss these in the workshop and identify improvement opportunities. For further analysis of interesting practices, site visits or task groups may additionally be organised.

With the best practices identified, participants should be able to draft their own improvement action plan. The action plan can be quite different for each utility and needs to be prioritized, based on the contribution of the proposed actions to the strategic objectives of the utility and the cost/benefit ratio.

Benchmarking without improvement usually equals frustration. The implementation step is often overlooked but is crucial in finishing the job that was started at step 1. In order to implement the suggested improvement initiatives, senior utility management should approve the necessary changes and the necessary internal procedures should be followed to secure investments, organisational changes, etc.

Chapter 9 covers the details necessary related to improvement actions.

3.1.2.6 Review of improvement actions

After implementing improvement actions, the results should be assessed to review if the objectives have been reached. Usually, this is done in a following benchmarking exercise. In order for the benchmarking process to be complete, this needs to be documented and evaluated, including lessons learnt and new benchmarking needs. Closing the cycle provides essential information for preparing a new benchmarking effort.

Chapter 10 provides additional information on how to close the loop.

3.1.3 Project milestones

Milestones in a project help to set clear targets for all those participating. Meeting deadlines in a benchmarking project is a key issue and a clear definition of milestones will undoubtedly help to that. In the present manual, project milestones will be presented according to the benchmarking process defined in this chapter (Figure 3.3).

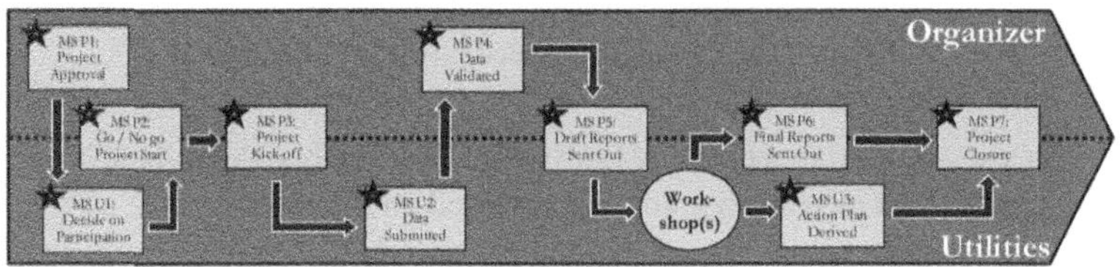

Figure 3.3 Benchmarking project milestones

Chapter 4

Prerequisites to successful benchmarking

Joining a benchmarking program without an adequate preparation implies the risk of facing disappointment. Despite the fact that experience shows that all participants can get some sort of benefit out of their first benchmarking experience, the amount of resources and efforts that are needed may not be compensated with enough performance improvement. This is usually the result of lack of determination, poor data quality and of willingness to share. In any case, any potential participants should make sure by reading this chapter that they are ready and willing to do what it takes to successfully benchmark.

This chapter will also provide guidance for organizers of benchmarking programs by advicing them on the key issues that can determine the success or failure of a benchmarking initiative.

4.1 HOW TO SUCCESSFULLY JOIN A BENCHMARKING PROGRAM

Despite the common preconception, joining a benchmarking program is not easy and is not just an economic matter. A serious benchmarking effort is the result of combining the work and interests of many important players, and in joining such a club, caution should be exercised. A failure to deliver what is expected from a utility in a benchmarking program will influence the perception other participants have from this organization. Additionally, a failure to successfully finish a benchmarking participation will be seen from the inside as a waste of resources.

This however should not discourage utilities from benchmarking. These are the key factors in order to succeed at a first attempt at benchmarking:

4.1.1 Alignment with strategic objectives

Benchmarking aims at continuously improving performance by learning from peers and adapting best practices. When utilities just join a benchmarking program because they are curious about their position or because individuals feel the need to do so, there is a risk of getting few or no benefits.

Just like performance assessment systems, benchmarking should be aligned with the strategic objectives of a utility, like providing a reliable service, reducing leakage levels or optimizing business processes to cut costs and to lower charges. By connecting strategic objectives to benchmarking activities, involved utility staff will be able to focus on the key issues and better prioritize identified improvement actions. Additionally, all resources used in the benchmarking program will be easier to justify if they contribute to reaching those strategic goals.

4.1.2 Commitment of senior management

Joining a benchmarking program requires full commitment from senior management to ensure that adequate resources are assigned and serious action plans are made to close the identified performance gaps. The commitment should preferably go beyond a single exercise, since performance improvement is a continuous process.

Once top management has decided to participate, the required resources need to be assigned. Skilled staff needs to be involved to timely submit data of good quality. Furthermore, funding needs to be arranged to cover project costs like staff, participation fees and travelling to attend meetings and events.

4.1.3 Willingness to provide good quality data in time

The performance assessment part is the foundation of any benchmarking exercise. Therefore, a prerequisite for successful benchmarking is the willingness of participants to provide the required quality data in time. It is essential that all participants can rely on the comparisons, and dubious data will invalidate all performance assessment results. The credibility of a benchmarking program depends largely on data quality, and the other participants will not accept (at least they should not accept) data not meeting the required quality standards.

4.1.4 Willingness to share knowledge and experience

Benchmarking is about sharing. In a competitive world, how much is shared is important and participants in a benchmarking project are not expected to share everything. However, the success of the performance improvement part of the exercise fully depends on the willingness of participants to share their knowledge and experiences.

For this same reason, involved utility representatives need to have an open mind when discussing results and ways for improvement. Many things can be learned from almost any other organization, provided one has an open mind to this. A proper benchmarking program will include a code of conduct that will clearly regulate how much information needs to be shared, with whom, when and how.

4.1.5 Stable organization

Benchmarking only makes sense when an organization is relatively stable and ready for further development and improvement. In case of utilities in a process of restructuring or merging, performance figures will probably show disturbances and will not be representative of the longer term trends needed to obtain reliable conclusions about position and possible performance gaps.

However, if the utility is able to deliver data from the time before restructuring, a before-after comparison can be executed to evaluate the outcomes of the change. Moreover, utilities confronted with change may consider joining a benchmarking program to identify practices that could guide them towards the desired new situation.

4.1.6 Benchmark at the appropriate level

Benchmarking programs vary a lot regarding scope and level of detail. Although one can benefit from almost every kind of benchmarking activity, the best results are obtained by utilities that join a program that fits their level of development, resources and availability of data. A program that is too simple for a utility may not provide any added value in terms of improvement opportunities. On the other hand, a program that is too sophisticated might be frustrating for participants that are not able to provide adequate information or have trouble interpreting the results.

As a general rule of thumb, smaller utilities are recommended to start simple and advance as they develop and get more experienced.

4.2 HOW TO SUCCESFULLY ORGANIZE A BENCHMARKING PROGRAM

4.2.1 Experience

An important prerequisite for the organization of a successful benchmarking program is the experience from the organizer. In the absence of experienced staff, the potential organizers of a benchmarking program should seek this experience in external individuals or organizations, at least from a supervisory point of view.

Benchmarking is a resource hungry activity (famelic, actually) for all those involved in it. In the absence of a professional attitude towards the organization of the program, or the experience of previous projects, all those resources may be wasted and the experience may turn into a disaster. The next chapters will provide further insight on the role of the organizer.

4.2.2 Comparable participants

Participants of a benchmarking program expect proper comparisons and being able to learn from each other. Ideally, this requires partners of comparable characteristics. Utilities which are similar in size, context and circumstances. However, in practice those partners are almost impossible to find, especially in the same benchmarking program. Each utility is unique: no one shares the same size of the service area, customer diversity, geography, type of water resources, treatment technologies, infrastructure, receiving waters, ownership, organization, etc.

For some utilities this is an argument not to engage in benchmarking efforts. After all, some organizations (like regulators) may feel the need to compare performance but why should utilities? Although proper comparison is the foundation of a benchmarking exercise, benchmarking is not about producing perfect comparisons or ranking lists. The use of performance measures in benchmarking is aimed to identifying performance gaps, best practices and improvement opportunities. Experience has shown that these objectives can be achieved even without the "perfect" comparison.

This does not mean that the organizer should not strive to find well comparable partners and – in the analysis stage – cluster utilities (for instance by scale, network structure or type of water resources) to facilitate the interpretation of results and the achievement of the objectives defined above.

4.2.3 Performance assessment system

At the start of a benchmarking program a decision needs to be made about how performance will be assessed. The performance assessment model describes the areas to analyze, the level of detail, and the performance assessment system chosen (performance indicators, variables and context information).

The IWA manuals of best practices on performance indicators (for water and wastewater services) represent a fantastic starting point for anyone designing a performance assessment system. The manuals have been tested in practice and are globally recognized as a standard in the water industry. By resorting to these manuals, the resulting system will share a similar structure and terminology with many other projects around the world. The indicators, variables and context information items found in the manuals provide a repository from which to choose and obtain inspiration. The thourough definitions guarantee that the wheel is not being reinvented, and if adopted, will contribute towards global standardization of definitions of variables and indicators, which will enable better exchange of knowledge and experiences in this area. A list of the indicators, variables and context information recommended by IWA can be found in Annex C.

4.2.4 Continuous process

Performance improvement is a process that never ends. Therefore, a benchmarking program should offer the possibility of repeated participation in specific benchmarking projects. The frequency of such projects depends on factors like the development stage of the participants, the diversity of topics of interest and the availability of resources. Typically, benchmarking efforts are periodically programmed with frequencies ranging from one to three years.

The organization of a continuous program requires some form of longer term commitment from participants and/or financiers. In the end, the continuity of the program will ultimately depend on the added value it generates for the participants.

4.2.5 Code of conduct

Benchmarking requires a few leaps of faith from participants, and since the stakes are high, the rules of the game need to be clear to everyone involved. Effective benchmarking requires professional behaviour from all players regarding organization, communication, willingness to share information, confidentiality, etc. These rules of the game, describing what is expected from both participants and organizers are clearly described in a protocol or code of conduct.

A good example is the Benchmarking Code of Conduct from the European Foundation for Quality Management[1] (EFQM). This code describes a set of basic

---

[1]See Annex B, chapter E.

benchmarking rules to ensure that confidential information stays within the group of participants and participants are willing to actively share experiences with the group.

4.2.6 Funding

Last but not least, a vital condition for a successful benchmarking program is obtaining the necessary economic resources for the project. Benchmarking costs include, but are not limited to, costs of staff, external assistance or consultancy, travel and accommodation, ICT, manuals, communication, etc. There are costs related both to the organization of the program as well as to the individual participation of each utility. Organizers should ensure that funds are secured for the project to be completed as planned. These costs may be covered by participation fees and/or external funding.

Chapter 5

Project planning

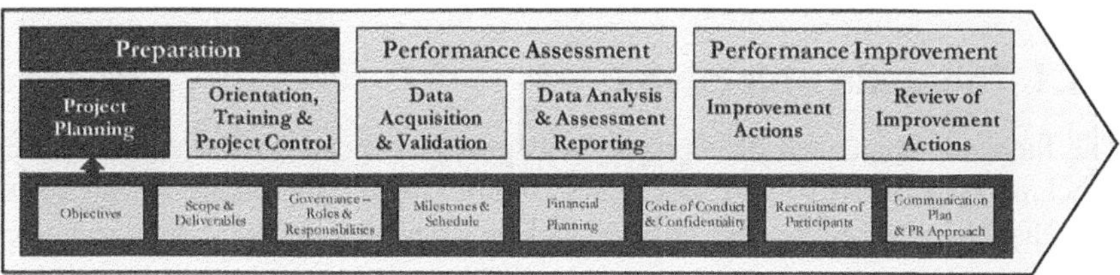

Basic decisions at the project planning phase can have a substantial influence on the design of the subsequent project steps and therefore on the success of the whole benchmarking project. Even from this early stage, it is necessary to consider the needs and areas of interest of participating utilities following a participatory approach.

The project planning phase includes two milestones on the project level and one milestone on the utility's side. Milestone P1 is the approval of the project including the required steps of setting up the thematic and organizational framework. Once the invitation with the project documents have been disseminated to water utilities, potential participants need to decide on their participation (Milestone U1). Sometimes, utilities can choose between two options: to simply participate in the benchmarking exercise as a benchmarked reference or to play an active role during the exercise (e.g. by being involved in an expert group). When enough utilities of similar structure have signed in, the go/no-go decision (Milestone P2) closes the project planning phase.

Within the next pages, the specific planning steps are discussed in more detail, both from the project and the utility perspective.

5.1 OBJECTIVES

The definition of benchmarking provided in chapter 1 clearly stated that the primary objective of a benchmarking program is to achieve the improvement in performance of the participating utilities. Additionally, benchmarking can fulfill other secondary objectives, like providing stakeholders with insight in the utility performance, and thus offering additional transparency.

However, before a benchmarking project is drafted, it is fundamental to clearly define the specific objectives and characteristics of the project. The benchmarking framework introduced at the beginning of this manual showed that benchmarking can focus at different levels and be applied at very different types of utilities. Therefore, it is critical to lay down the main objectives of a project for all participants to see what they are up to.

5.1.1 Thematic objectives

The thematic objectives of a project will greatly depend on the scope of the benchmarking activity. The objectives should reflect the areas of interests and challenges faced by the utilities and be aligned with their own strategic objectives.

Moreover, the objectives shall be well-balanced in order to gain performance pictures meeting the manifold demands in reality. For example, it is recommended to consider both cost efficiency objectives and service quality objectives (e.g. reliability of the service).

5.1.2 Methodological objectives. The "triple C"

In addition to well defined thematic objectives, a benchmarking program should pursue certain general methodological objectives to ensure its quality and value to the participants. These objectives can be summarized by the "triple C":

- Comparability and high data quality throughout the benchmarking process.
- Continuity. The benchmarking strategy should be considered on the long term and not be focused on a single project.
- Compatibility with international standard systems for cross-border or cross-program referencing and not inventing the wheel again.

5.1.3 Individual utility objectives

Utility

In general, different utilities will have different objectives, which should obviously be compatible between them and compatible with project objectives. Each participating utility should list its own objectives and expectations for the benchmarking project, at least cross-check them with the project objectives and communicate them to the project group in order to integrate them in the project setup.

Possible utility objectives in a benchmarking activity include:

Systematic acquisition of essential utility management data
- Establishing or improving measurements and reporting environment.
- Involving employees in utility data handling (training, personnel involvement).

Assessment of current performance
- Assessment of performance relative to certain levels of service and/or targets.
- Assessment of relative position with respect to other utilities in a similar operating environment.

Assessment of performance trends (periodic benchmarking)
- Evolution of performance compared to certain levels of service and utility targets.
- Evolution of performance compared to other utilities in a similar operating environment.

Design and review of improvement actions
- Exchange of knowledge and experience with other utilities.
- Identification of performance gaps (where to improve).
- Adoption of best practices at process and task level (how to improve).
- Evaluation of improvement actions' results by repeated participation.

Transparency to utility owners and stakeholders
- Public accountability of utility performance.
- Providing evidence of need for investment plans and priorities.
- Showing willingness to improve.

5.2 SCOPE AND DELIVERABLES

As hinted earlier, a sound project scoping is crucial for the establishment of the performance assessment model, its data requirements and the project deliverables. In addition to organizational matters of project governance, the project scope represents the basis for the decision of utilities on whether to join a particular benchmarking effort or not.

5.2.1 Target group of water utilities

A more or less homogeneous group of utilities (or a high number of participants clustered into peer groups of comparable utilities) is the key to gain comparable results. Typical sorting criteria are:

- Geographical scope (regional, national, international).
- Type of activity
  - Water supply (whole, retail and/or bulk supply).
  - Wastewater (sewerage and/or sewage treatment).
  - Other activities (e.g. electricity, gas, telecommunication) for cross-sector benchmarking.
- Utility structure (e.g. network density, water and wastewater treatment requirements).
- Utility size (especially for adapting the extent of the benchmarking effort).

5.2.2 Scope of performance assessment and improvement

In order to define the scope of the benchmarking project, utilities should be surveyed about their areas of interest and challenges regarding their own internal improvement (strategic objectives). Additionally, information about their need to communicate their performance status and trends to their stakeholders may also be relevant to determine the final scope.

Another key issue is the level of detail for the project (utility, function, process or task, see Figure 1.1). This should include:

- Defining well-balanced performance categories related to the chosen level of detail (see examples in Annex B, chapter B)
- Defining assessment criteria related to the performance categories, in order to align the later chosen performance assessment system (see recommended procedure in ISO guideline series 24500 in Annex B, chapter B).

Finally, the extent of jointly organized performance improvement activities should be determined (workshops focused on interpretation of results, utility visits for exchange of best practice knowledge, workshops on improvement action planning etc.).

5.2.3 Deliverables

Depending on the scope of the project, the following deliverables can result from the project:

- **Data questionnaire and IT solution** for performance data handling.
- **Training workshop and documentation.**
- **Utility visits for data validation,** by the project team.
- **Validated individual data**.
- **Utility Individual Report**, draft and final, only assessment results or including improvement action plans.
- **Consortium Report**, draft and final (often combined with individual reports).
- **Public Report**, draft and final.
- **Workshops** and workshop minutes (on cause analysis of assessment results, on best practice exchange, on improvement action planning).
- **Individual presentations, workshops and deeper analyses and consulting**.

5.2.4 Scope and deliverables for utilities

Utility

A water utility willing to participate in a benchmarking project should answer the following questions prior to entering a benchmarking project:
- **Does my utility fit into the target group of utilities?**
  - If not, is there a chance to integrate a group of similar utilities to the project?
  - If yes, are similarly structured utilities willing to participate as well?
- **Does the thematic scope cover my areas of interest and the challenges my utility is facing?**
  - If not, can I actively influence the drafted scope to be amended to my demands?
  - If yes, can I expect useful answers from the level of detail of the applied benchmarking model? (It is recommended for newly-participating water utilities to start at high-level – utility level – to obtain a good overview and to go into more detail with further exercises).
  - If yes, can we afford the efforts for the planned level of detail and is the expected cost/benefit ratio good?
- **Can I expect benefits from the planned deliverables?**
  - Do the deliverables meet my requirements (e.g. usability of IT solution and questionnaire, quality of illustration of results and readability of reports).
  - Which deliverables are within the project frame, which are optional extras?
  - To which extent do I want to be supported/involved in jointly planning of improvement measures?

(Continued)

> **Utility** (*Continued*)
>
> Large utilities can also resort to carrying out internal benchmarking activities. In such case, some of the key issues raised in this scope and deliverables section may be answered internally (both on the utility and organizer role).
>
> These may be tackled by identical departments of different locations, related to utility-wide performance measures. Or they could also be arranged cross-sectorally (electricity, gas, telecommunication, etc), comparing PIs and/ or practices.

5.3 GOVERNANCE – DEFINING ROLES AND RESPONSIBILITIES

Defining roles and responsibilities within a clear project structure is a key step towards an efficient and reliable project atmosphere. Especially during the establishment of a new benchmarking activity, the different actors and their rights and duties must be clearly specified in order to gain mutual trust in the benchmarking effort.

5.3.1 Who is benchmarking?

The actors participating in a performance assessment effort, and even their roles from one effort to the next one, can change depending on what motivates the initiative. For instance, it will depend on whether the assessment activity is launched "bottom-up" by the industry (e.g. by national water associations or several utilities) or "top-down" (regulatory authorities or funding agencies recommending or requesting utilities to deliver the data for a performance assessment effort).

However, it should be noted that these latter activities should better be called comparative performance assessment, yardstick competition etc. and not benchmarking, since the innovative element of learning from each other (the performance improvement phase) is usually missing.

The following setup of players and accountability structure has proven to work well in industry-driven benchmarking projects and could be considered a typical cast of actors in a benchmarking project:

5.3.1.1 Project responsible body

Typical tasks of the responsible body (organizer, owner, initiator) of a benchmarking effort are coordinating the participants, commissioning the project team and surveying the benchmarking process. A project responsible body is often a consortium of water utilities or a national association.

5.3.1.2 Project steering group

Additionally to the responsible body, a project steering group may be established. The steering group is often composed by representatives of the project responsible body. It controls the project development through the benchmarking process. It is sometimes divided into a larger strategic committee and a smaller project group.

Additionally, stakeholders like authorities (sometimes funding the activities) can be invited to a larger steering committee in order to involve them in the general project development.

5.3.1.3 Participants

Participants in the benchmarking projects covered by this manual are usually water utilities except in those cases where cross-sector benchmarking also takes place (in such cases, organizations from other sectors, typically utilities, can also participate in the project).

5.3.1.4 Operational project team

External benchmarking experts (consultants, academics, etc.) can also be commissioned to provide additional expertise and to cover the whole range of performance areas (technical, environmental, financial and organizational) leading through the operational tasks of the benchmarking process.

As several benchmarking programs are run by the industry itself, often project teams are created including employees of the participating utilities.

Examples for program governance and project structures are given in Annex B, chapter C.

5.3.1.5 Utility benchmarking roles

Utility

Participating water utilities usually act as the "customers" of the benchmarking effort, although sometimes they are also involved in an expert advisory group or even as part of the project responsible body and the project steering group. The potential roles of a utility in a project are:

- **Participant**
  Entering into contractual arrangement for the benchmarking exercise as a benchmarked object, providing expectations and general utility information for project planning purposes.
- **Active member in an expert advisory group or even in the project steering group**
  Playing an active role in project planning (e.g. having influence on project scoping) or during the exercise.

5.4 PROJECT SCHEDULE

Benchmarking activities are often founded upon figures of the annual report and annual accounts. In those cases, it is only after those figures are consolidated that the performance assessment phase can start. Since project assessment results should be available within the subsequent year for planning and control reasons, the schedule and duration of the project is one of the critical success factors.

Experience shows that projects are prone to delays at certain critical points in the process. These critical stages are fundamental to determine the project milestones as shown in chapter 3. More specifically:

- **Getting enough participants before the scheduled project start**
  Utilities need time to internally communicate the activity and to decide on the participation. Projects require a critical number of participants to be able to proceed. Consequently, the earlier the project announcement is placed and potential participants are contacted, the better the time schedule can be achieved.

- **Time needed for data acquisition & validation**
  Especially in large utilities, data acquisition is divided between several divisions and merged before the approval. This usually leads to a lengthier process that can delay the delivery and validation of data. The project team should urge participants to meet project deadlines. In these situations, it usually pays off to do so personally or by phone.
  Additionally, in larger projects, additional personnel resources should be allocated to the project team to allow simultaneous utility visits for data validation.

- **From draft reports to final reports**
  Often in a project, while some utilities have already started their individual analyses on the draft report, others are still waiting for the approved final version and facing a time delay. Therefore, an additional milestone should mark the time when the feedback to the draft reports is submitted by all utilities.

In some benchmarking projects, the performance improvement phase can become a task that is left to utilities to pursue or at least to peer groups of utilities, especially when a large number of participants joined the performance assessment phase. In these cases, the schedule for this phase becomes a more individualized task. Even when this happens, the official project closure can be arranged by means of a review workshop and feedback from the participants (e.g. from a questionnaire) should at least be integrated in a final project document.

Examples for a project schedule are given in Annex B, chapter D.

> **Utility**
>
> Water utilities willing to participate in the benchmarking effort need to make sure at an early stage if they can cope with the scheduled deadlines and ambitious time slots. Quite often, underestimating the extent of data requirements leads to a delayed start in data compilation.
>
> Meeting deadlines often depends on the initial conditions of data availability, and whether data can be quickly compiled or not. Beginners often find out that they have less data than they originally thought they had, that sometimes data are available but harder to collect than in the initial expectations or that the quality of these data is poorer than expected. In the end, it may be a matter of assigning and coordinating sufficient personnel resources in order to get "fresh" results.

5.5 FINANCIAL PLANNING

5.5.1 Costs

Benchmarking can be expensive. Usually, participants cover the costs of their own activities themselves. Additionally, a fixed fee is usually charged to cover common project costs.

Depending on the project design, costs can result from:

- **Costs of participant activities**
  - Personnel hours
  - Travel and accommodation costs for utilities (e.g. workshop and meetings participation).
- **Common project costs:**
  - Personnel hours and travel and accommodation costs from the project responsible body.
  - Assistance of the project co-ordinator by an external facilitator/ project team (e.g. costs for deriving the data questionnaire, for data validation, compilation and analysis, for reporting, for facilitating the improvement phase).
  - Organization of the kick-off meeting and the workshop.
  - Organization of company visits (during data validation or for know-how exchange during the improvement phase).
  - ICT costs (like website, database, certificates, licenses, etc.).
  - Translation costs, printing, mail.

The common project costs can be broken down in direct and indirect costs. Direct costs are related to each participant (like individual data validation) whereas indirect costs are independent from the number of participating utilities.

5.5.2 Covering costs

Benchmarking activities generally provide the opportunity to share costs among the participating utilities so that added value can be achieved at reasonable price, but it obviously depends on the number of participants. The more utilities joining the effort, the better the ratio of sharing indirect costs is (general project management, setting up the performance assessment system, public reporting and communication). Usually the final number of participants is not fixed during cost planning, and therefore the offered participation fees must be based on assumptions.

In any case, price setting, cost planning and the number of utilities form a linked system which the specific strategy from the project responsible body must take into account. Criteria of such a strategy can be:

- **Achieving representativeness for the water sector**
  The target of this strategy would be to maximize the number of participants while regarding their belonging to different categories of water services. Sufficient participants from each category would facilitate clustering into different peer groups allowing both for comparability for each utility within the peer groups and for getting a sector picture of different utility categories as well. This strategy can be promoted by different measures: graduation of fees according to utility size, offering rather low prices (e.g. by co-funding of projects from public organisms or other organisms).

- **Achieving high individual benefits for participating water utilities**
  The targets of high quality standards and individual customization generally implicate higher project costs and consequently higher fees for each utility. In any case, it is up to the project team to achieve efficient overall procedures to allocate budget for individual deliverables like management summaries with specific recommendations in individual reports. Finally the success of the project will mainly be assessed by the individual benefits for each water services.

- **Splitting participation fees in modules**
  This strategy provides the opportunity to keep the prices low for newcomers offering a basic module ("benchmarking light") with a leaner assessment system at lower costs, but also with fewer results. The answer to how can a lean a system still be beneficial for the participants, depends on the objectives and hence the scope of the benchmarking effort.

In some cases, a tariff splitting between a basic fee for the performance assessment phase and optional extras for the performance improvement phase (e.g. workshops) is sometimes applied. This implies flexibility for the utilities, but also some disadvantages: worse planning reliability at project level and the risk that participants consider their benchmarking efforts finished after the performance assessment and do not continue with improvement steps.

Utility

Utilities have to consider the following cost components when taking part in a benchmarking activity:

- Participation fee to cover the common project costs.
- Cost of the IT solution provided by the project.
- Fees for optional extras like additional workshops, presentations, company visits, analyses and additional consultancy.
- Costs for own personnel resources including travel expenses.

The last component is especially important: once the results of the performance assessment phase are available, work in the utility to start transforming them into improvement actions begins. And this work needs personnel resources to be assigned!

Funding of utility costs is usually self-covered. Sometimes direct public funding of utilities is offered by water authorities, mostly to trigger first time participations.

In fact, utilities base their decision on participating upon the question of whether the benchmarking effort pays off in terms of a cost-benefit analysis, an answer which cannot be given individually at an early stage, when the decision is made. Experiences show that once performance gaps are recognized, closing them usually produces much higher benefit than the resources needed to detect them.

5.6 TERMS & CONDITIONS: CODE OF CONDUCT AND CONFIDENTIALITY REQUIREMENTS

Benchmarking efforts are highly affected by what we could consider "psychological factors" originating in the participating utilities and their personnel as their performances are compared and assessed. After all, utility performance is the responsibility of its staff: from the top-management to the blue-collar workers. However, it also depends on the context in which the water service is embedded in, the behaviour of shareholders and on the decisions and actions of former staff, due to the long-lasting effects within water services.

In a benchmarking project, it is crucial to achieve an atmosphere in which sensitive data, especially weaknesses, are handled with care. A clear code of conduct helps accomplishing trust within participants and a pleasant atmosphere of collaboration which is beneficial to everybody involved. Ultimately, the feeling of participants competing against each another needs to be avoided at all costs.

For all these reasons, new projects are usually set up with rather strict confidentiality arrangements. Although this is necessary for launching the activity, it also prevents from deriving deeper results from the comparing and learning process. The greater the barriers of confidentiality are, the more difficult it is to visualize results and their causes, and the knowledge exchange is greatly affected.

An important task of the project team is to keep an eye on the psychological momentum and to create conditions which all of the participants agree to. For example, the confidentiality rules might be loosened during workshops when orally discussing the assessment results.

Generally, three spheres of confidentiality requirements could be distinguished:

- **Interaction between participant and project team**
  The project team needs a good insight into utility data. Consequently, the project team should get as much support as necessary to carry out the assessment and to assist the utility during performance improvement. This obviously implies that the project team must guarantee the highest degree of confidentiality.

- **Interaction between participant and participant**
  A proper way to describe participant interaction would be *"everything is possible – nothing is a must"*. Provided that information on a third participant will not be disclosed, participants can flexibly arrange their exchange of data, results, knowledge and experience.

- **Transparency to the public**
  Depending on the project characteristics, the nature of the information given to the public can range from just some general project information to the delivery of named performance values or rankings. Examples of data masked or blinded by a coding system are given in chapter 8.3.

An example for a code of conduct from the European Foundation for Quality Management as well as an example for a confidentiality agreement are shown in Annex B, chapter E.

Utility

Water utilities should actively claim and contribute to clear project conditions, as they will definitely influence the quality of the collaboration.

However, when joining a well-established project, utilities should also acknowledge that they are expected to adhere to project conditions.

5.7 RECRUITMENT OF PARTICIPANTS

Once the project plan is approved by the project responsible body, the main contents of the project documents need to be transferred into different communication media like brochures or project webpage in order to allow the utilities to get a quick view on the proposed benchmarking exercise.

Good marketing practices, like a coherent corporate design, can help to promote the proposed benchmarking activity. Nevertheless, it is important to cover the project contents from the utility perspective and to answer the questions that utility representatives will likely raise when deciding on participation.

5.7.1 Communication channels

Recruiting the maximum number of participants within industry-driven projects should not be considered as the ultimate objective. As experience shows, it is mostly the utilities that have been inherently motivated for participation that will benchmark in a proper way and will achieve high quality data.

Consequently, benchmarking is a matter of self-conviction and not of being persuaded by someone else. Utilities need genuine information on what benchmarking can accomplish and which barriers and limitations have to be considered.

This is why we named this section "communication channels" rather than "marketing your project", for these communications efforts should seek the self-persuasion mentioned above:

- **Letter from the project responsible body or a well known third party (e.g. associations)**

  A more than adequate starting point of the recruitment phase, but not a stand alone policy. A letter (or e-mail) might be sufficient to mobilize utilities already experienced and convinced of benchmarking. However, although new utilities may obtain some basic information from the letter, they will need more details for their decision.

  Common topics to be clearly covered in a project invitation letter are:
  - The objectives of the exercise.
  - The addressed target group.
  - The expected benefits for participating utilities.
  - The scope of the exercise.
  - The actors in the exercise.
  - The performance assessment model.
  - The terms of confidentiality and data processing.
  - The project deliverables and interactions (reports, workshops, company visits etc.).

- The participation fee, additional expenses and expected personnel involvement.
- The timetable (e.g. including project steps and milestones).
- Reference projects and success stories.

Examples for the invitation of potential participants are given in Annex B, chapter F.

- **Personal contact by members of the project group**
  Despite being a very time-consuming channel (the effort should not be underestimated) it is useful to reinforce the information on the project that utility managers have. Contact can be arranged by means of company visits, meetings at conferences and exhibitions or at least phone calls, the latter usually as a follow-up and reminder of previous communications.

- **Word-of-mouth advertising**
  The more recognizable a project is as an industry-driven benchmarking effort, the larger the impact communication between utilities will have. This communication channel only works adequately when the general atmosphere between participating utilities is trusty. Another option to minimize the efforts is to get into networks (e.g. of water associations, of public utilities and municipalities) to make use of the momentum of trusted communities.

- **Additional on demand information**
  It is recommended to offer additional information on demand (e.g. on a project website) but this cannot substitute triggering actions like personal contacts or letters.

Besides the several channels of communication, the choice of the addressee needs to be considered. A choice that is not easy. Which rung of the hierarchy ladder shall be contacted?

It is recommended to first address top management (and also shareholder representatives in publicly owned utilities, e.g. mayors), because they are responsible for the performance of the utility and its continuous improvement.

Of course, operational managers and staff must be convinced of the effort as they will be responsible to deliver good quality data and to implement the improvement actions afterwards. Thus, general managers should involve the operational staff by convincing them of the importance of performance improvement and the usefulness of benchmarking.

5.8 COMMUNICATION PLAN & PUBLIC RELATIONS

Especially for benchmarking activities that involve many different players and interested parties, a clear communication is essential for both the successful execution of the project and the acceptance from the stakeholders. Therefore, internal and external components of project communication can generally be distinguished when deriving a comprehensive communication plan.

On the one hand, an effective communication within the benchmarking exercise allows for an efficient and transparent project. This requires an adequate intensity of information exchange, as well as the prevention of misunderstandings and extra work to be done due to missing information.

On the other hand, benchmarking efforts should not be seen as actions limited by topic or a time framework. The communication regarding the benchmarking project and its outcomes to the external stakeholders is fundamental to get them involved in the world of benchmarking. Once stakeholders consider benchmarking as an essential instrument for better meeting their needs, the sustainability of the benchmarking effort is affirmed, and so is the relationship between the water utilities and their stakeholders (Figure 5.1).

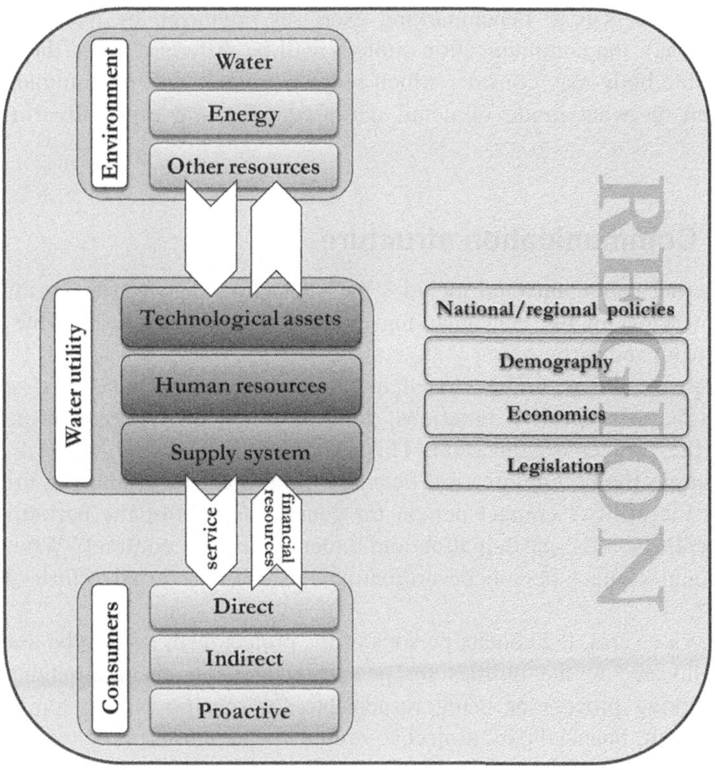

Figure 5.1 The water utility context (Source: Alegre *et al.*, 2006)

Preparing a distinct communication plan in advance of the exercise is a useful practice that includes:

- Defining the communication strategy and objectives.
- Defining the communication structure within the project.
- Defining the communication actions to be included in the project plan.

5.8.1 Communication strategy and objectives

The communication strategy will be developed in close connection with the objectives of the benchmarking exercise. When the exercise only aims at the performance improvement of the utilities (e.g. within a project which is solely organized by utilities), the communication strategy will mainly focus on utility-internal workflows and on the information of their board of directors and of the shareholders.

If the exercise has also got the objective of documenting transparency to the stakeholders (e.g. benchmarking exercises organized by national water associations), the communication strategy will be different. Then, the project responsible body may consider which project contents will be communicated to whom, at what grade of detail and frequency and especially for what reason.

5.8.2 Communication structure

The internal communication includes the interactions between and within the different actors in the benchmarking exercise, like project responsible body, project team and participants.

Moreover, it has to be considered that each actor consists of several persons having specific functions, both in their own organizations and within the project organization. This should be considered when defining for example the communication between the project team and the utilities: Who is the utility's contact person for general aspects of the participation, like deciding on the participation and undersigning the contract? Who is the operational contact person co-ordinating the utility-internal efforts on the project?

And, vice versa, the contact persons of the project team need to be assigned, being allocated to the utilities for personally assisting them throughout the benchmarking process or being responsible for specific project topics (e.g. technical part, financial part, project co-ordination/administration).

5.8.3 Communication actions

According to the chosen communication strategy and structure, the specific communication actions can be planned in more detail:

- Which interactions at which project step (e.g. recruitment of participants, staff training within the project team, team-internal quality control sessions, training of the participants, help desk in data acquisition, data transmission, data validation, communication and discussion of assessment results, improvement action planning etc.).
- Which channels are to be used for collaboration (personal workshop, company visits, e-mail, phone, project website, newsletters etc.).
- Which channels are to be used for documentation (project reports, press conferences, articles in journals and newspapers of public administrations, presentations at scientific conferences and industry conventions, newsletters, project website, weblogs etc.).

Examples for newsletters and project articles are given in Annex B, chapter G.

Utility

Water utilities can contribute to the success of the project and their participation by communicating right from the beginning their willingness to improve. The earlier they get their employees and stakeholders informed, the easier it will be to report the process step-by-step and therefore to implement improvements by the benchmarking effort. Utilities can make the employees and stakeholders prepared in time by informing about the objectives of the benchmarking efforts, about the benchmarking process and about a mixture of good and not so good assessment results, the latter to be utilized for knowing where and how to further improve.

Water utilities can then play an active and self-confident role in the communication of benchmarking outcomes with their stakeholders and the public. Possible activities could be in particular:

- **Communication at the local level of the served area**
  Utilities can transparently report about their experiences gained from benchmarking efforts, their benefits and improvement actions derived. Utilized channels could be individual presentations to the municipal representatives, articles in local newspapers etc.
- **Communication at the project level**
  It is essential that the outcomes from benchmarking exercises are communicated from the perspective of the utilities. Recommendations from

(Continued)

Utility *(Continued)*

"customers" are much more powerful than other promotional activities! For instance, utilities can present case studies at conferences or journals, with lessons learned.

- **Internal communication to employees**
In order to utilize the assessment results for improvement actions, it is important to get the employees on board early because it is them that make things happen in the end. Similarly to other strategic issues within the company, benchmarking should be seen as a positive initiative by the utility staff in order create change.

Chapter 6

Orientation, training and project control

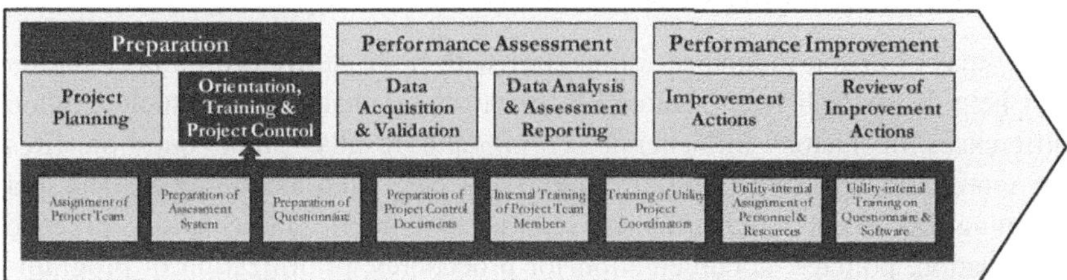

Participants in a benchmarking project share different backgrounds and levels of expertise. Critical elements of any benchmarking project are the project and utility specific training and quality control that will be required for successful implementation and execution. In this chapter, typical elements associated with both project and utility level training and quality control that would be necessary are presented.

6.1 PROJECT LEVEL OBJECTIVES

6.1.1 Project plan & protocols

Project group members, including the project steering committee, sponsoring organization, third party consultants or overall project managers, technical advisors, and lead coordinators from each utility are among the stakeholders that would be invited to participate in project training. Such training and orientation is often conducted in a setting attended by this entire group, or major elements thereof. It may last for 1–3 days.

Examples for agendas of orientation and training workshops are shown in Annex B, chapter H.

Orientation and training follows the project planning phase, and would incorporate the key elements and objectives, assessment process, analysis and reports, workshop plans, and protocols to be followed. In addition, detail would be provided on process and schedule, resources to be used and required, roles and responsibilities, introductions of key staff and participants, survey or questionnaire content and guidelines, software training, web support and help facility, protocols for communication and problem resolution, and any contextual differences from past benchmarking processes.

Annex B, chapter I serves with an example of an assessment guidance document.

6.1.2 Data questionnaire

Data surveys or questionnaires often start with basic utility profile information that help distinguish type, size and other attributes that will ultimately explain differences or factors influencing scoring and results. Some questionnaires for more detailed projects require additional data such as strategic plans and planning documents, performance reports and metric information, practice information, policies, documentation for procedures, prioritization of programs or environmental factors, governance structure, customer base, and the like.

6.1.3 Software, tools and web support

Sometimes participants and third party facilitators or reviewers or coordinators have not used the particular software being employed or are not familiar with the framework, process or content of subject matter being addressed. Processes for managing the data and its integrity need to be explained. Explanations about how data accuracy and consistency will be achieved. Also, benchmarking projects will usually include some form of user help facility and web based support. This needs to be explained and demonstrated at this project training.

An example for an acquisition software manual is given in Annex B, chapter I.

6.1.4 Project kick-off

It is usually best to conduct a project kick-off meeting at the project level as a way to maximize information transfer most efficiently, provide a common frame of reference, assure consistency in communication and put all utility coordinators and others on the same footing. Standard elements of a kick of meeting would be:

- a published agenda
- introduce project organization, roles and responsibilities
- cover basic content and process
- considerable time for questions and interaction.

Sometimes it is useful for past participants in benchmarking projects to describe their experiences and lessons learnt to new participants, including how they managed past efforts, internal processes and staff engagement, as well as how the benefits of benchmarking have been applied in their local setting.

6.1.5 Other factors to consider

There are many examples of effective benchmarking project organizations, practices and training. Below are some additional aspects to consider as training and orientation is provided:

- **Review Methodology and Project Plans**
    - Outline of benchmarking objectives and process.
    - Strategy for ensuring objectivity (replication) of the outcomes.
        - Strategy for managing risk of systematic bias in assessments.
        - Overview of the data acquisition and data validation process, schedule, contact details, communication channels and protocols, help desk and key personnel.
- **Quality control and management process**
    In the following table, a series of quality control measures for a benchmarking project is presented. This establishes a number of procedural steps to assure data integrity, consistency and quality across the range of participants.
- **Validation and data management:**
    Data validation process is fundamental in the project. There are several issues and expectations, from the consultant and the participants, that need to be taken into account:
    - Consultant or third party reviewer will have access to key supporting information.
    - On relatively minor issues, the consultant may accept utility advice regarding availability of documentation or process.
    - Wherever practical, the consultant should cite relevant documentation and/or field notes.
    - Consultant or third party reviewer should document where there is a significant issue or data questions.

Table 6.1 Quality management of the benchmarking project

Phase	Quality Management Task
Project Planning	• Project Plan: Scope, methodology, governance, roles and responsibilities, schedule, confidentiality, communications, audit protocols, conflicts of interest • Team planning workshop • Steering Committee approval of report templates, questionnaire & interview guide. • Reviewer accreditation and training • Team selection/pairing and rotation
Orientation	• Onsite Interview schedule confirmation • Orientation workshop presentation review
Data Acquisition & Validation	• Review team briefing: Review/interview processes, consistency, objectivity • Communications feedback and FAQ update • Data management • Any utility self assessment and questionnaire reviews • Structured interview agenda confirmation • Standard interview material development and review • Team teleconference on consistency and objectivity • Overall consistency review
Analysis and Assessment Reporting	• Draft Utility reports; conduct team review or quality control • Public report assembly, documenting overall project, including any outside Advisory Team review • Draft Public Report and sample Utility Report review
Final Reporting	• Final Report reviews
Best Practices Workshop	• Best practice utility selection and review • Best Practices Conference presentation review
Post Project	• Post implementation Review

- **Professional objectivity**

  The consultant should pursue several objectives during the course of the project regarding its outcome:
  - Objectivity: e.g. to what extent may the process or data be subject to interpretation)
  - Replication: if the current practitioner(s) were to leave, would the practice or level of performance be clear to a qualified person from outside the utility

- Materiality: do the responses differ in a material way?
- Staff knowledge (is the matter fully and consistently understood by staff).

- **Statutory Obligations**

  Typically in a project (especially in an international one) different participants are subject to varying regulatory environments. In some cases, the participants may have already been subject to audit on specific statutory issues. However, it should be clear that outcomes of a regulatory audit do not necessarily imply conformity with the specific benchmarking process being undertaken.

- **Disputes**

  Disputes occur if there is a disagreement between participant and the third party assessor and there is a risk that differences between views would result in a material difference in the overall assessment, or if there is an issue of conflict of interest or professional integrity.

  In such cases, a brief summary of such disagreement will be documented and an established dispute resolution procedure will be followed.

6.2 UTILITY LEVEL OBJECTIVES

Utility

Participants in the project need a reasonable level of management and planning to efficiently achieve the best value and outcomes from their self-assessment – as with any project.

6.2.1 Roles and responsibilities. Internal coordinator and team

After orientation on the project as a whole, specific preparation within the utility needs to occur in order to appropriately resource the effort. A project coordinator is normally selected to manage utility specific responsibility. Depending on the scale and complexity of the work involved, one or more teams are generally assigned to assemble data, collect profile and survey information, assess practices, perform quality control functions, prepare documents, participate in interviews, review findings and recommendations, mobilize additional staff as needed, and to serve as advocates and even trainers for organization resources.

These roles should not only be identified, but communicated broadly to the utility, along with statements of roles and responsibilities and expectations. Sometimes this can be accomplished following an official chartering exercise, where a clear mission, goals, deliverables, schedules, budgets, and outcomes are specified.

(Continued)

Utility *(Continued)*

6.2.2 Schedule
A detailed schedule of activities needs to be developed that will meet all the requirements of the project as a whole, including survey and assessment tasks, production of deliverables, reviews and feedback, and formulation of improvement requirements.

6.2.3 Assignment of resources
Benchmarking projects are usually a priority for utilities; and to achieve necessary quality for this effort, it is necessary to assure competent and respected individuals be assigned to perform this activity. For credibility purposes, as well as to provide valuable information exchange, a highly capable individual should be designated as coordinator, and a broad based group of individuals should be assembled to represent the various functions and subject matter being explored.

6.2.4 Internal training on questionnaire and software
Depending on the benchmarking project involved, as well as the form of involvement by different internal parties, training will be necessary for staff in order to assure a smooth and efficient assembly of data in the appropriate form for analysis. Roles and responsibilities should be identified for data input, performing any analytical and validation processes, interpretation of results, formulation of any special information required, and potentially in-sourcing the software for future use in the utility.

6.2.5 Helpful hints
Following are training and orientation process steps that reflect lessons learned from past benchmarking projects that may prove useful to utilities engaging in similar efforts. The steps indicated below will need to be adapted for individual circumstances.

Assign a credible and reliable utility project coordinator
- Coordinate the various activities and resources within the utility.
- Communicate with management and obtain their strong commitment on the Project to assure organizational alignment during the entire process.
- Assure that key managers understand the objectives of the project, consider expectations for them and their staff, required input, key schedule dates, and authority and responsibility relationships among participants.
- Provide answers to key questions and quality control to ensure consistent interpretation of the framework.
- Provide a key point of contact with project and outside parties.
- Generally speaking, the coordinator's role should NOT include data collection or validation (at least without additional help).

(Continued)

Utility *(Continued)*

Start early
- Document objectives and expectations of the project.
- Allocate all necessary resources required to complete the data collection and interpretive work.
- Be realistic about schedule and allow sufficient time for resources to complete assignments, while honouring project commitments.
- Identify and track key milestones.
- Attempt to measure what the utility gains from benchmarking, including business process and performance improvements.

Establish clear roles and responsibilities
The responsibility for each step of the benchmarking process, and the role that each actor is playing should be clearly outlined and communicated.

Determine the realistic level of effort that is required
Consider there is a direct relationship between the amount of effort in management, planning and resourcing the project and the learning to be gained. Be mindful that the level of effort required will be quite variable among the participants, largely because of the scale and complexity of the utility.

Plan for resources
Plan for resource commitments that will be required for attendance at orientation and software training; compiling and delivering input data; assessing and preparing supporting documentation; conducting any onsite interviews; review of reports; attendance at assessment and best practices workshops; selection and documentation of improvement actions and lessons learned at project completion.

Surveys and questionnaires must produce objective and verifiable data
They should produce credible scoring. But this requires care and an understanding of the information. Justification and documentation are critical to both the utility participant and any reviewers. Documentation should be available to support scores.

Attendance at internal training and orientations should be mandatory
Likewise, it is critical for attendance at best practice workshops, a venue that offers the greatest learning opportunity, along with utility networking.

Chapter 7

Data acquisition and validation

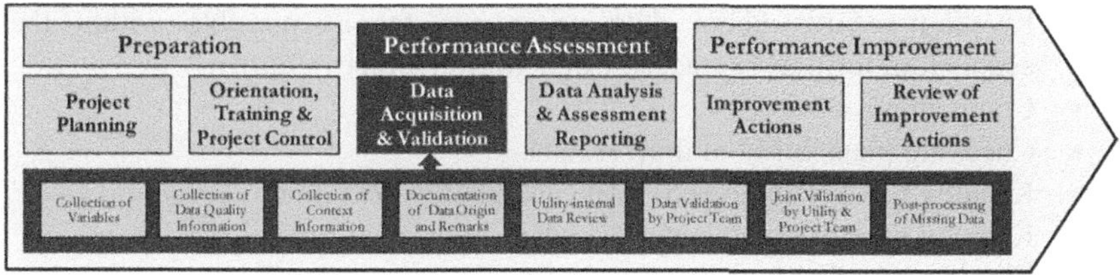

7.1 COMPARABLE DATA – THE LINCHPIN FOR SUCCESSFUL BENCHMARKING ACTIVITIES

Gaining comparable results during the performance assessment phase is essential for benchmarking efforts. Misleading conclusions might be drawn during the performance improvement phase when the famous "apples and oranges" are compared.

This counts for any statistical comparisons, but especially for performance assessments where actions are systematically derived from comparison results and even more for regulatory activities where actions are implemented top-down.

Nevertheless, procedures have been developed in practice to assure good data quality and comparability – or better to reduce the influence of still existing differences as complete comparability can never be achieved with reasonable efforts while gaining traceable results.

As it can be seen from the following list of factors influencing data quality, comparability is not just restricted to the project step of data acquisition, but is crucial throughout the whole benchmarking process. For instance, the first two factors are the general basis that must be established during earlier project steps.

Otherwise, failures in comparability cannot be eliminated even by high efforts during data acquisition!

- Capability and willingness of personnel to deliver high quality data:
  - Voluntary and anonymous participation.
  - Training.
  - Assignment of adequate personnel resources (qualification, time budgets).
  - Co-ordinating internal team.
- Clear and precise definitions in data element system.
- Clear data questionnaire (incorporating both usability and comprehensive data definitions as well).
- Validity checks in questionnaire (to avoid misunderstandings and mistakes just during the data acquisition).
- Keeping records about data origin (as far as possible within the acquisition form to be traceable also for the project team).
- Company visits.
- Checking plausibility of draft results.
- Keeping records about better data acquisition next time.
- Keeping records about better data definitions next time.
- Elimination of bad quality data and/or
- Visualization of inaccuracies in performance charts.
- Comparability of single PIs between utilities (influence of denominator choice).
- Comparability of single PIs over time (dynamic change of denominators).

Most of the previous factors are covered in detail within the next sub-chapters.

7.2 DATA ACQUISITION – TRANSLATING REALITY INTO FIGURES AND FACTS

7.2.1 The questionnaire

Data acquisition starts as soon as the questionnaires are sent out to the participants or the connectivity to web-based acquisition forms is activated. In practice, this is usually achieved either by spreadsheet files or a web based application.

Spreadsheets are more flexible, allow for easy programming and are powerful in combining additional calculations, validity remarks, graphically supported data forms and visualisation of draft charts. Standard spreadsheet software is common to everybody, although different software versions and configuration may cause some interference. Unless properly developed, spreadsheet files are prone to undesired manipulation from participants, and submittal and processing must usually be done manually.

Web-based applications have the advantage of direct communication between participant and project database, and do not require any software installed (except for a web browser). The project team can track the acquisition progress from intermediate data savings and urge if necessary. A clear workflow (clearing of data forms, handling of older versions and traceability of submissions) is facilitated by web-based data exchange. The same website used for submitting data can be used to publish results and to enhance project communication.

Special performance assessment software is available on the market combining the flexibility of customizing the PI set and stand-alone applicability with web-based communication facilities between participants and project team, and some even provide the option of ready to use custom spreadsheets.

In any case, the choice of data acquisition model used should respond to the needs and reality of the participating utilities. In the case of countries with limited resources, internet connection speeds and the existence and capacity of computers should be taken into account as limiting factors.

Regardless of which way data is chosen, the following generic requirements have to be met:

- **Comprehensive description of required data**

  Translating the developed performance assessment system into the data questionnaire is not that simple as it occurs at first sight. It is absolutely necessary to have all the definitions easily available during the acquisition process, for all engaged persons, at any time and directly within or close to the questionnaire. On the other hand, utility personnel must not get lost in plenty of details, but keep a clear view on the acquisition progress. E.g. good experiences have been made with pop-up windows serving with definitions and additional remarks. Anyway, it should be taken into account that such temporal windows do not occur in printed versions.

- **Usability**

  Data quality can be seen as a function of data availability on the one hand and of the capability and spendable time of the utility personnel on the other hand. Especially the latter is essentially influenced by the handling quality of the questionnaire and its IT solution. Features enhancing usability could be pull-down menus for selections of given categories, graphically supported data forms to view the context of the data, referencing to technical standards etc.

- **Validity checks**
  It is recommended to incorporate validity checks right within the questionnaire visible to the persons filling in the form. This reduces misunderstandings and mistakes just during the data acquisition. Such validity checks could be checksums of water or energy balance data, of financial balance or personnel data. Remarks could appear when a certain presumed data range or a certain data format is not met.

- **Confidence grading**
  Additionally to the required data, meta-data are needed for qualifying their robustness during the analysis. The IWA system of performance indicators includes a confidence grading system (Alegre *et al.*, 2006) which can be used or amended for any performance assessment project (see chapter 2.1). It is also viable for essential data elements (e.g. water balance data) to incorporate deviations of measuring devices directly into calculations gaining PI values together with inaccuracy bands. And the reliability of metered water data can easily be screened by the question when the meters have been calibrated for the latest time. An example is given in Figure 8.3 of chapter 8.3.2.

- **Space for data derivation and remarks**
  Keeping records about data origin and about data amendments during the validation process should be done (as far as possible) within the acquisition form. This is necessary for tracking the data derivation when the performance assessment results are interpreted and when the data shall be derived in the same manner during upcoming performance assessments. These meta-data are also important for the project team validating the delivered data whether the data have been derived homogeneously by all participating utilities.

7.2.2 The data collection step

It is the duty of the project team to support the utilities during their data compilation, both on the level of the data themselves and on the workflow of data acquisition as well. While different communication channels can be applied (e-mailing, panel in the project website, personal assistance, etc.) phone calls have been chosen by the utilities most frequently getting information from bi-directional communication in short time during the acquisition process.

Whereas oral information is fastest and unbureaucratic, the quality of information has to be assured within the project team. It is highly recommended to keep records on the consultations, to talk them over within the team and especially to document the problems participants are facing in order to be able to incorporate them in the questionnaire for the next project.

Furthermore, the team is responsible to monitor the participating water utilities if they are keeping to the time schedule and, if required, to remind them. Thereby, phone calls proved to be more powerful than sending e-mails.

Utility

The first task during data compilation is to achieve an overview on the required data by an assigned core person or core team. It is warmly recommended to start with the initial steps immediately after receiving the questionnaire. These initial steps are:
- Screening required data volume and data quality.
- Identifying data sources.
- Identifying data gaps between requirements and availability, and
- Allocating the data collection to the different divisions.

Especially when involving further employees into the data collection process, they must get informed about the purpose of the benchmarking exercise. If they are not motivated or even anxious about what could happen with the data, they will not elaborate the data with due diligence. Moreover, time targets should be arranged to keep on schedule.

Data collection & recording

One of the consequences often occurring during data collection is the localization of data gaps or the cognition and adoption of better data recording structures facilitated by the benchmarking form. Therefore, data collection efforts comprise a learning process, not only the improvement actions based on the assessment results. Optimally, the data recording system within the utility is adapted to the performance assessment system so that benchmarking data can be derived in the future by just several mouse clicks (in fact, the assessment system should generally be adapted first to the strategic objectives of the utilities and accordingly to their existing office automation systems installed in practice).

Adopting the benchmarking system for utility-internal recording depends on the fact whether the gathered data can be utilized also for other purposes,

(Continued)

Utility *(Continued)*

like planning (e.g. asset management plans, water safety plans), strategic and operational controlling, reporting to authorities etc.

Anyhow, an increased effort for data collection has to be regarded especially for the first in a benchmarking activity. A crucial element is the documentation of the data sources. Recording data derivation (as the interface between existing data structures and the required benchmarking data) assures the traceability of the benchmarking results and their re-embedding into the utility-internal structure.

Often, the question of how to cope with unavailable data is raised. If the questionnaire allows for quoting data reliability and accuracy (and it should), a best possible assumption shall be given anyway. By erasing the data gap until the next benchmarking activity, the quality of the assumptions can be checked and give interesting hints on the necessity of data recording.

Internal data review & validation
Before submitting the data to the project team (milestone U2), an internal data validation loop facilitate an already high data quality. Criteria of the internal data check are
- Regarding a high quality data submission (the product):
  - Completeness of data and required meta-data (confidence data, additional remarks).
  - Consistency of data (with internal records, to required definitions).
  - Completeness of meta-data regarding data derivation.
- Regarding the workflow of data collection (the internal process)
  - Identified data gaps and better data recording structures.
  - Identified organizational weaknesses (e.g. unclear responsibilities).
  - Identified psychological effects (exposure to assessment activities: how do we cope with being evaluated?).

7.3 DATA VALIDATION – QUALITY FIRST!

7.3.1 Necessity of data validation

Once more, it must be highlighted that it is essential to validate data prior to the data analysis and draft-reporting step. Experience shows that (despite best efforts in data collection) misunderstandings and mistakes occur often. Individual data might also be of best reliability and accuracy, but fail to fit into the requirement of homogeneous data collection due to special circumstances.

But, how can the project team validate the data?

7.3.2 Back office validation

At first, incoming utility data shall be checked by the project team using several methods:

7.3.2.1 Individual plausibility check of draft PIs

A simple procedure is to check the preliminary PIs if they are lying within the expected range or not. This first plausibility check can also be embedded directly in the questionnaire by automatically colouring all PI values not fitting into the expected range.

7.3.2.2 Crosschecks of variables

The consistency of the submitted data can be checked on the level of variables, e.g. by looking at the relation of the number of employees to the personnel costs and getting a more or less plausible average value of the man year.

7.3.2.3 Outlier analysis

As soon as enough data sets are received from the utilities, statistical calculations of outlier selection can be carried out. Furthermore, it is recommended to interpret the preliminary PI results. Bar charts and scatter plots (without the need of anonymity in the back office) can be quickly drawn to get a rough idea of the relative position of the participants.

However, outlier analysis alone may not be sufficient: e.g. a PI value of poor data quality which is among the best performing, but does not appear as an outlier, will not be filtered by these methods. Especially for the very important variables (like water balance/waste water balance, finance data, personnel figures) data derivation shall be talked over together with utility representatives.

Utility

Joint validation by the project team and the utility
Especially for utilities participating for the first time in a benchmarking exercise, a visit of a member of the project team is of indispensable importance. The central aim is checking the data together, the utility and the benchmarking expert, by structured interviews to grant the required data quality by looking at the data origin and data derivation. Once the benchmarking expert is acquainted with the individual context, he/she can validate data levels

(Continued)

> **Utility** *(Continued)*
>
> much better. Also the draft assessment results and first opportunities for improvement can roughly be discussed during the on-site bilateral data workshop.
>
> Company visits are also recommended for subsequent projects and at most can be substituted by longer phone calls. Within new projects, it is necessary to discuss issues on the background of the benchmarking process and its contents in order to build confidence in benchmarking and in the relationship between utility and team.
>
> An example for an on-site interview protocol for data validation is given in Annex B, chapter J.

7.3.3 End of validation step

After data review and post-processing (if needed) are finished, a clear cut shall be communicated by reaching the milestone P4 (data validated). Submitting data amendments at later project steps only be accepted as additional remarks. Otherwise they would delay the forthcoming data analysis step.

Chapter 8

Data analysis & assessment reporting

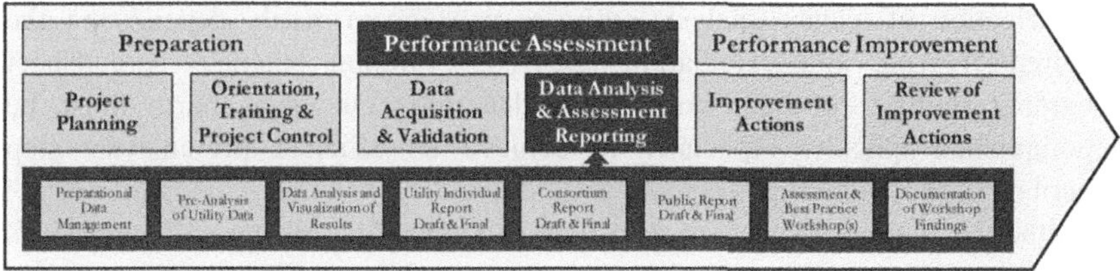

8.1 SYNTHESIZING DATA

At this stage, the project team has validated data in the database (in case of a web-based data collection solution), or in multiple spreadsheets which need to be compiled in one spreadsheet (or, preferably, one database).

Consequently, some prearrangement of data is necessary before the performance comparisons can be accomplished. Furthermore, data analysis software must be selected and tested based on the choice of which graphical charts and calculations shall be achieved. As the data structure is already known, these preparation steps should be carried out by the project team in advance to save project time.

8.1.1 The need for a database

Historically, a significant number of performance assessment and benchmarking projects have used spreadsheets to collect and process project data. While former spreadsheet formats were restricted to a certain number of columns (e.g. 255 columns) new versions have become rather powerful and are capable of working with larger data volumes.

However, spreadsheets are not specifically designed to achieve complex queries and relationships between data, and present certain difficulties when data need to be exported to other data analysis software. These shortcomings can somehow be overcome with a significant amount of programming and the use of macros, and often are.

However, the need for a database becomes evident as soon as more than one benchmarking cycle has been carried out. Then at least two values, two confidence values and two remarks for each data element for each water utility must be stored and queried in one consistent data environment without redundancies – a typical case for relational data bases.

8.1.2 Data analysis software

The selection of a data analysis software package primarily depends on the required charting and calculation features. Common business spreadsheet programs might not cope with these demands. Drawing charts must be accomplished quickly, especially to achieve an efficient pre-analysis step described in the following.

Although there are software packages encompassing all the software needs for a benchmarking project (web based/spreadsheet data collection, data storage and data analysis) this is not the common rule. Exceptionally, some larger projects have software designed from scratch specifically targeting their actual needs.

8.2 PRE-ANALYSIS

Before the performance comparisons will be carried out, preliminary analyses are recommended to get a "feeling for the figures". This is especially true for projects with very demanding confidentiality requirements where single results cannot be charted or interpreted. These tasks are left to the project team in the back office. Depending on the experience of the project team, the heterogeneity of the participants and the novelty of the scope, this will be a more or less elaborate effort. However, the resources needed for this step should not be underestimated, especially in new projects.

Pre-analysis can focus on the following topics:
- Analysing the profiles of the participating water services and clustering them according to main explanatory factors for performance differences.

- Sensitivity analysis of clustering criteria for different performance measures.
- Sensitivity analysis of different denominators for determining which PIs would fit best to assess performance in a certain area.
- Comparability of single PIs over time (dynamic change of denominators).
- Additional utilisation of performance indices and multivariate data models (e.g. frontier methods, data envelopment analysis) to gain overall performance hints (for advanced analysis in the back office).

8.2.1 Clustering the utilities

The linchpin of comparable assessment results is the appropriate clustering of utilities into rather homogeneous peer groups according to different kinds of influencing factors. Data elements like context information, primary variables like utility size (e.g. displayed in m^3 system input or km network length or pollution load in inhabitant equivalents), aggregated index values (e.g. summarizing the extent of organisational provisions) or even interrelated PIs (e.g. PI leakage control for water loss PIs) can be used as explanatory factors.

Typical clustering criteria for instance can be found among IWA standardized context information elements for water supply (Alegre *et al.*, 2006):
- Network delivery rate (m^3/km.year) (IWA code CI76).
- Service connection density (No./km) (IWA code CI61).
- Consumption per service connection (m^3/conn.year) (IWA code CI72).
- Origins of water sources (IWA CI95 to CI100) and type of water treatment (IWA CI27 to CI30: no treatment/disinfection only/conventional treatment/advanced treatment).
- Type of asset ownership (IWA CI3; categories public/private/mixed)
- Type of operation (IWA code CI4; categories: corporation/municipal/ associations...).

In addition to one-dimensional classifications of the utilities for each of the clustering criteria, statistical methods like cluster analysis can be applied to gain multi-variate peer groups. For instance, the above mentioned network delivery rate (CI76), service connection density (CI61) and consumption per service connection (CI72) can be combined into a grouping criterion describing the urbanity of the water supply service (metropolitan/urban/rural).

8.2.2 Sensitivity analysis of clustering criteria

As the central reporting aim should focus on simplicity, the most influencing explanatory factor should be used as the clustering criteria in the reports.

In addition to expert opinion (or common sense), statistical methods like regression and correlation analysis and graphical visualisations are useful tools. For instance, scatter plots and grouped box plot charts (see chapter 8.3) help to filter the most influencing factors.

8.2.3 Sensitivity analysis of different PI denominators

In each performance area (e.g. reliability or efficiency), performance assessment criteria are present in the numerators used to calculate the different indicators. For instance, personnel costs usually cover personnel efficiency from the financial perspective. However, personnel costs can also be related to different denominators (e.g. to the supplied/treated water, network length, number of service connections, etc).

Depending on the denominator used for the performance assessment, the relative position of a particular utility compared to its peers may change significantly. Special attention should therefore be paid to key performance criteria like cost efficiency and personnel efficiency. As the numerators total costs, running and capital costs, and number of employees are highly aggregated data elements, the choice of the denominator in the PI formula is of crucial importance when analyzing results.

The PI values in Figure 8.1 are derived from the Austrian "Stage C" project. All five utilities belong to the same group of network delivery rate (scales are not shown due to confidentiality of the project, but are not relevant to the point under discussion). Analysing the positions of the utilities, utility 29 would be the most inefficient one according to the "total costs per volume of water supplied" (vertical axis). However, 29 is in the middle of the pack when "total costs per mains length" is considered (horizontal axis). Moreover, utilities 31 and 36 rank quite low in costs per m^3, but lead the pack in costs per km. Finally, 15 is last in both categories, but when a third indicator (costs per service connection) is considered, 15 is no longer last (15 has a network with low density of service connections).

Figure 8.1 is the demonstration that depending on the selection of indicators, and even the selection of denominators, the result of a comparative performance assessment may be completely different. Needless to say that this is a key issue in which participants would certainly like to have a say. For this very same reason, the selection of PIs and their denominators should be undertaken systematically with clear criteria (established in advance) and the analysis of results should be based on the simultaneous study of kPIs with different denominators to get a comprehensive picture of the assessed criterion or performance area.

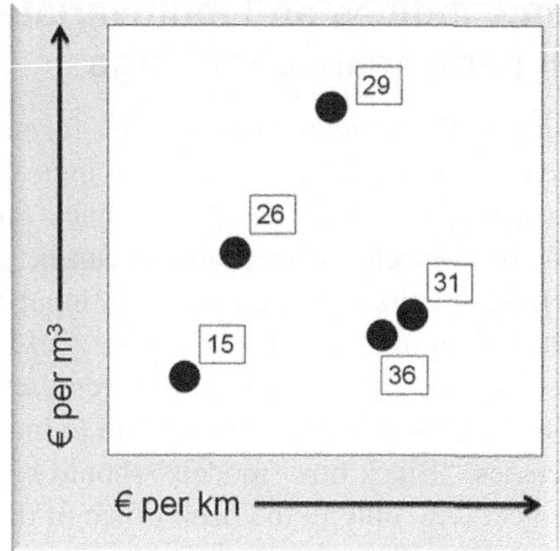

Figure 8.1 Scatter plot of the total costs per delivered volume versus total costs per mains length for five utilities (Theuretzbacher-Fritz *et al.*, 2009)

8.2.4 Comparability of single PIs over time

Besides comparability issues between utilities within one benchmarking project, comparability of single PIs of different years (trend analysis) allows to determine how each utility has developed over time compared to other utilities.

When trends are analysed for a group of utilities, additional considerations need to be taken into account. First, it has to be assured that only those utilities that took part in each benchmarking are included in the comparison. Comparing different utility samples will most probably lead to misinterpretations.

The assessment criteria are still the numerators in the PI formulas, but in trend analysis these are considered as they develop over time. Unfortunately, not only the numerators are varying temporally, but also the denominators. And even worse, so are the relative differences between denominators from different utilities.

It has to be accepted that each utility has its own specific story which will affect both the assessed criteria and the denominators of the indicators (and therefore the reference to which indicators are made relative to).

In benchmarking projects at the utility level, this kind of analyses provides valuable insight in the strategic behaviour of each utility. In these cases, workshops among the utilities will be the best instrument to undertake time series analyses and to derive improvement action plans.

8.2.5 Performance indices and multivariate data models for overall performance rankings

In addition to assessing multiple scores or rankings for selected performance areas, an overall score can be calculated by weighing indicators and aggregating them into a performance index. Additionally, complex data analysis models (e.g. frontier methods, data envelopment analysis) can be used to gain overall performance hints. However, the selection of their input measures and their weightings can bias the output of the models (Cabrera *et al.*, 2009b).

Since the addressees of the analyses (the utility representatives) are usually not familiar with these methods it is difficult for them to incorporate them in the decision-making processes. "Black box" models[1] should be used very carefully in a benchmarking project (e.g. only in the back office of the project team or by knowledgeable experts). The risk is that these methods just lead to rankings of utility performances, without understanding where and how these utilities could improve.

8.3 DATA ANALYSIS

The more elaborate the pre-analysis has been, the faster and more goal-oriented that the actual performance comparison can be accomplished towards draft reporting. The choice of tables and graphs illustrating the comparison results mainly depends on the grade of confidentiality agreed upon within the project.

This section overviews the most common results analysis and charting options.

8.3.1 Tables

Tables show single values. Therefore, they are easy to understand and allow for a deeper analysis of results. However, it takes time to interpret figures, locating maximum and minimum values, finding correlations, etc.

The figure shows an example of the Dutch public benchmarking report (VEWIN, 2007). In the example, these disadvantages are reduced by adding different shades of grey to the cells (the darker the cells, the higher the values).

---

[1]The term makes reference to the mathematical difficulty to understand and follow what these methods actually do to process data and should not be taken as derogative in any way. However, it is a fact that to the average user, these powerful tools are less transparent and more "obscure" than simple ratios.

Table 8.1 Example of cost PIs (€/connection) visualized in a table (from the public report of the Dutch water supply benchmarking, VEWIN, 2007)

	Total costs	Taxes	Cost of capital	Deprecia-tion	Opera-tional costs
WBGR	167	29	11	20	107
Brabant Water	178	34	33	27	83
WMD	187	29	32	31	95
Vitens	187	31	34	33	89
Waternet	208	6	21	47	134
WML	211	24	62	40	85
DZH	213	19	49	49	95
PWN	219	4	33	50	132
Evides	219	10	87	47	76
Oasen	245	28	32	67	118
Sector	200	23	42	39	96

In benchmarking efforts with high confidentiality requirements, other chart options described later should be used instead of tables. Even if utility names are not shown, participants could be traced back by analyzing the different table values.

8.3.2 Bar, grouped bars and stacked-bars charts

In general, bar charts are easy to understand as they show a single value for each bar. Their application in benchmarking efforts goes from describing the general context of the participating water services (how many utilities of which structure, of which size…) to picturing performance measures when confidentiality issues are not very strict. As a side note, 3D bars should be avoided since it is usually difficult to determine the actual value represented by each bar.

Bar charts should also be avoided in confidential projects. Even when single bars are only identified by numbers and not by utility names, single performance measures may be traceable to uncover a utility, as it happens with tables.

Figure 8.2 shows a bar chart from the 2009 project of the European Benchmarking Cooperation. The indicator "total cost coverage ratio" shows to which extent total costs are covered by total revenues. The bars show the distribution of PI values of the participating utilities. Values of different years can easily be set side-by-side in bar charts showing also the development of each utility. The dashed horizontal lines visualize the average and median value of the analyzed group.

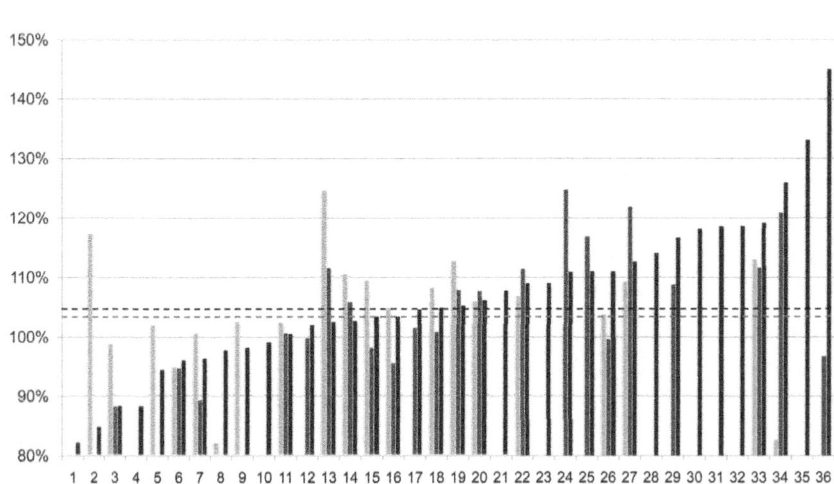

Figure 8.2 Example of a bar chart (from the European Benchmarking Cooperation EBC)

Figure 8.3 shows a bar chart comparing 11 participants of the Austrian benchmarking on the process of water loss management (Koelbl *et al.*, 2009). This and other charts were used as the starting point for analyzing performances and the applied methods. As a consequence, additional information was packed into this chart: urbanity as a crucial context information (bar colour) and the accuracy ranges (horizontal lines "min., max.").

The grouped bar chart allows presenting aggregated figures for each peer group (e.g. median values) and can facilitate a combined view on more than one performance measure. See for example Figure 8.4 where the shares of several cost components of the running costs are shown.

The stacked-bar chart can be used to show the accumulation of several inter-related figures according to each utility or to each peer group (e.g. Figure 8.5). If the stacked-bars are created from more than one data set (e.g. for a number of utilities within a peer group), the statistical function must allow the accumulation. Therefore, arithmetic averages were preferred to median values in the case of Figure 8.5 (shares of personnel with different qualification levels which must add up to 100% of total personnel).

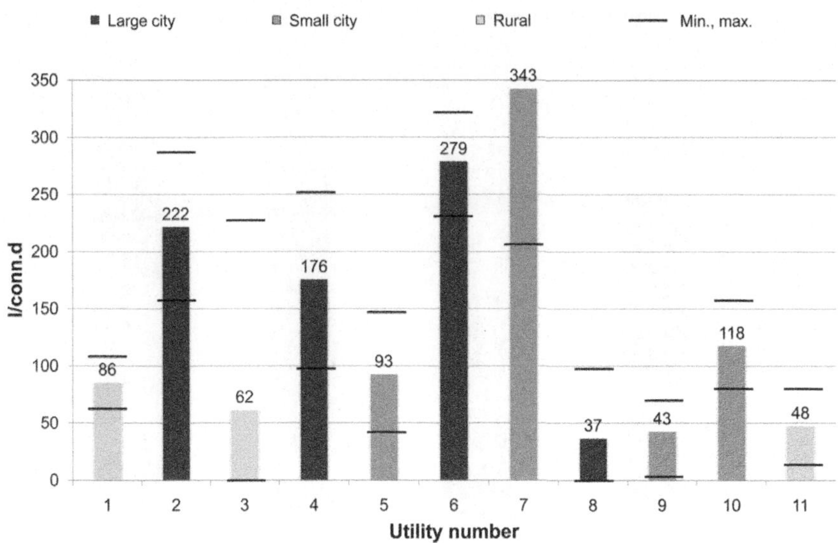

Figure 8.3 Bar chart with additional clustering and accuracy bands (Austrian water supply benchmarking)

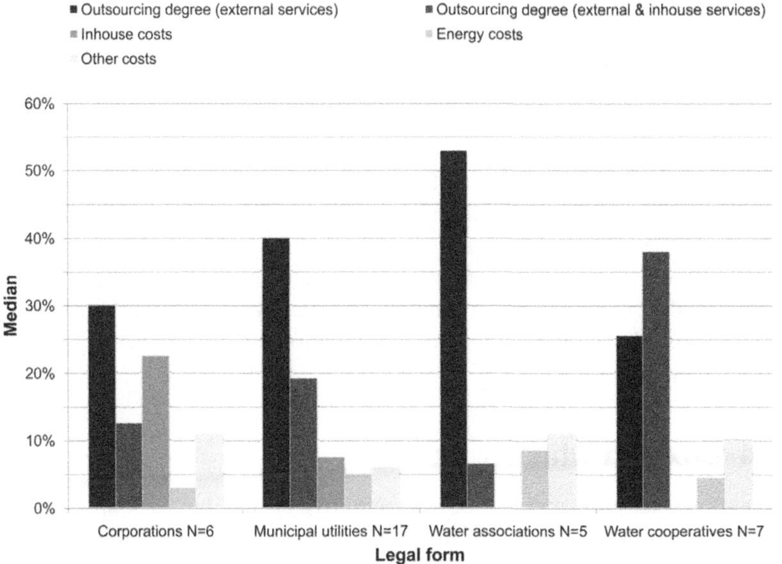

Figure 8.4 Grouped bar chart (Austrian water supply benchmarking)

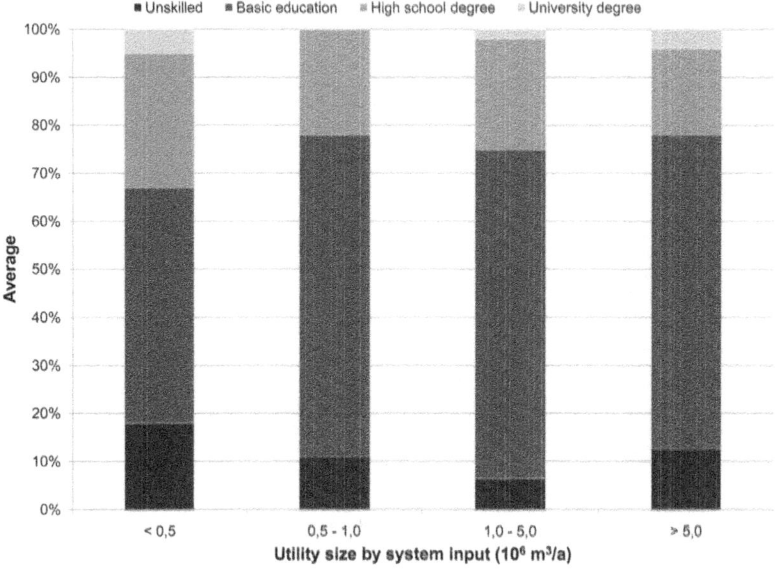

Figure 8.5 Example of a stacked-bar chart (Austrian water supply benchmarking)

8.3.3 Scatter plot

Scatter plots are the perfect tool when performance is the consequence of assessing the distribution of values over two axes. These can be the values of a performance indicator viewed together with a numeric explanatory factor or a combined view on the cross-relation between two performance measures.

An example of a scatter plot is provided in Figure 8.6. It shows the distribution of costs for customer metering and billing per meter (x axis) versus a quality index (covering a number of mostly cost-driving practice level queries). This allows for a combined interpretation both of efficiency issues and of service levels as well.

8.3.4 Box and whisker plots

Box and whisker plots may not be very intuitive at first, but as soon as the user is familiar with them, they allow for a quick interpretation of the distribution of PI values. Box and whisker plots also assure a high degree of confidentiality.

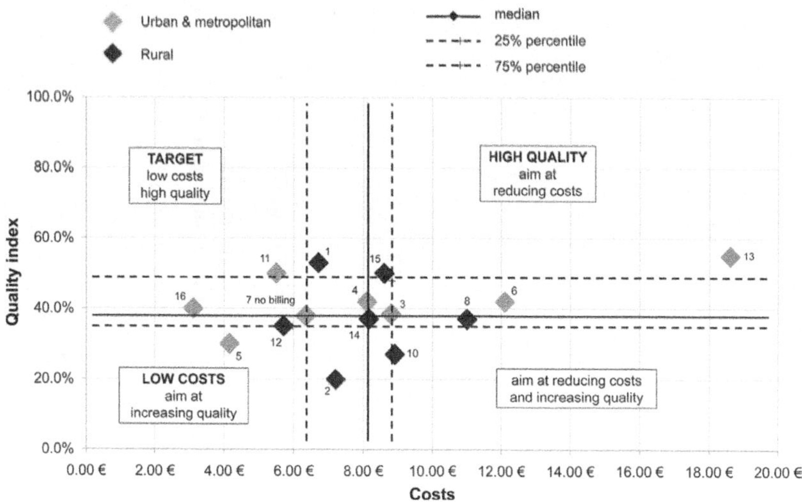

Figure 8.6 Example of a scatter plot (Austrian water supply benchmarking)

A box and whisker plot summarizes the main information for a group shown in a bar chart (e.g. Figure 8.7 left chart). More specifically, the minimum, maximum and median values of a distribution. The median[2], although not as common as the mean or arithmetic average, is not influenced by extreme values (outliers) which can disfigure the result.

The black t-ends of the box plot (the "whiskers") represent the minimum and maximum values (see Figure 8.7). The grey box shows the lower and upper quartile, showing where the middle half of the sample values is. This is a very simple way to spot the middle field of participating utilities (contained within the grey box boundaries). Moreover, outliers are pictured as extra dots when their values are at a certain distance from the rest of the sample (1.5 times of the box length from upper or lower quartile). The small white mark in the box indicates the number of valid values visualized in the plot.

Clustering into peer groups can be achieved by placing several box plots side by side in one chart. Experience shows that despite a steeper learning curve, once mastered the box and whisker plots represent a very powerful synthesis tool.

---

[2]The midpoint of the range numbers that are arranged in order of value.

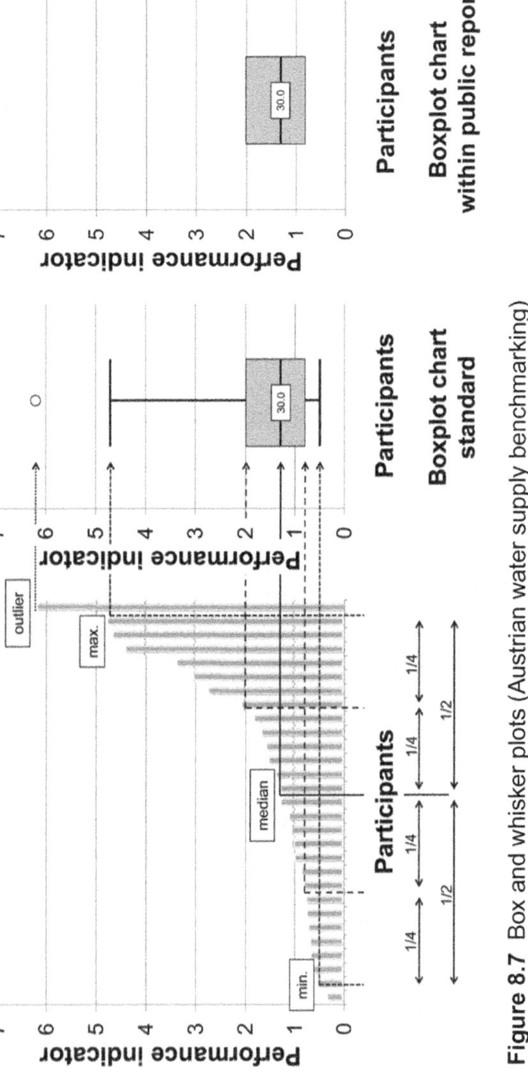

Figure 8.7 Box and whisker plots (Austrian water supply benchmarking)

8.3.5 The "reduced boxplot"

The chart on the right in Figure 8.7 visualizes how comparison results can be brought to the public by both facilitating sufficient transparency and granting confidentiality as well. As minimum and maximum values might result from specific circumstances of single utilities, an option is to simply display the middle 50% for public communication (a box with no whiskers).

8.3.6 Box plot chart, clustered in peer groups

Combining several box plots of different peer groups enables the analysis of the different "behaviours" of the groups. Usually participants will only compare themselves to the peers in their group. However, the differences between groups contain important information that shows why comparing non-clustered utilities or comparing performance without context information is hardly ever advisable.

Figure 8.8 shows the energy costs per m^3, clustered the required water pumping head. The higher the water has to be elevated, the higher the energy costs usually are per m^3. However, even then the dispersion of values can be quite high (due to differences in energy efficiency levels and energy tariff levels, and also due to the steep gradient of the unit energy costs over the pumping head).

In these cases, utilities with average pumping heads close to those of neighbouring peer are recommended to also take those groups into consideration.

Figure 8.9 takes the combination of box plot charts a little farther, showing a combined chart of two inter-related performance measures. The bright boxes cover the extent of tasks outsourced to external firms while the darker boxes show the extent of tasks outsourced both to external firms and in-house overheads as well.

From this clustered chart it can be easily concluded that different levels of personnel and outsourcing costs have to be expected for different legal frameworks for utilities[3].

Finally, Figure 8.10 shows yet another example of box plots (reduced). In this case, the time trend of twelve utilities from the Austrian water supply benchmarking (Neunteufel *et al.*, 2009) is visualized between the years 2004 and 2007. Additionally, the information of a single utility is also displayed. Both the group and the single utility have achieved progress regarding the index performance measure to customer relations.

---

[3]In corporations and municipal utilities, the water services are embedded in larger organizations (energy supply companies, municipal administration, etc). On the other hand, "water associations" (regional water companies in Austria serving and led by a number of municipalities) and local water co-operatives (villages) are organized as stand-alone organizations.

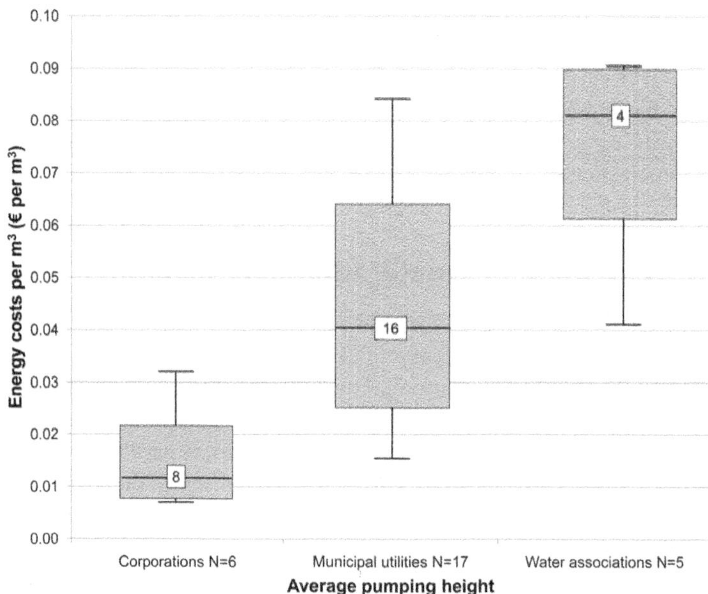

Figure 8.8 Box plot charts clustered in peer groups (Austrian water supply benchmarking)

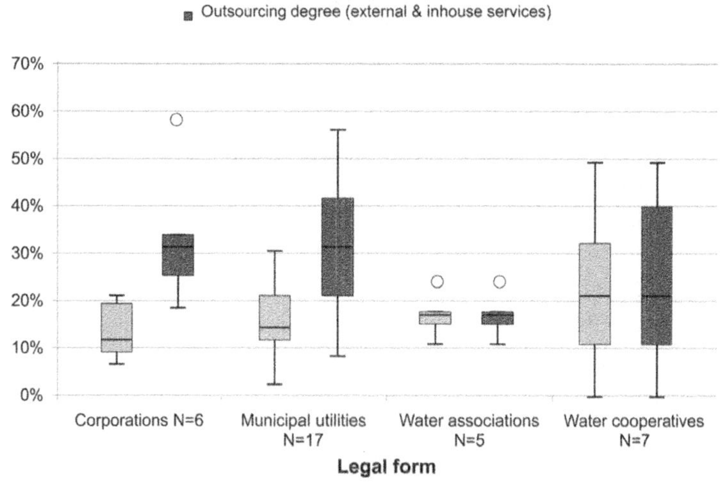

Figure 8.9 Clustered box plot chart for two different indicators (Austrian water supply benchmarking)

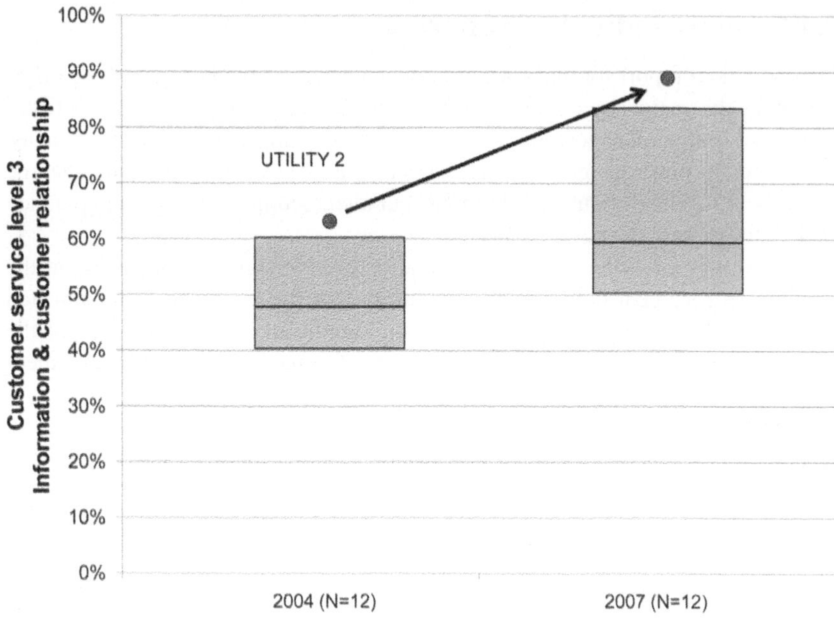

Figure 8.10 Time evolution of reduced box plots and single utility (Austrian water supply benchmarking)

8.4 REPORTING – DIFFERENT REPORTS SERVING DIFFERENT TARGET GROUPS

Obtaining the right answers from a benchmarking project is not an easy task. On a first effort it is common to find mistakes (e.g. at data collecting) or doubt about the comparability of certain figures. For this reason, it is essential to facilitate a test loop of draft reporting before the final reports are derived. After a first draft report participants may also complete the supplied information, allowing to refine the assessment results. These contributions could include specific circumstances not covered by the acquired data or the suggested reasons for the assessed performance. In any case, this exercise of finding performance drivers is one of the key issues at the assessment workshop taking place after the draft reports are issued (see next chapter).

There are several reports that usually result from a benchmarking project:
- Utility Individual Report. Draft report with assessment results and first recommendations, and a final one including improvement action plans.
- Consortium Report, draft and final (often combined with individual reports).
- Public Report, draft and final.

8.4.1 Utility individual report

The individual report for each utility is the core tool of the benchmarking effort regarding the main objective of utility improvement. It is developed to foster buy in by senior management and to facilitate the implementation of changes. Therefore, it must be concise and targeted, and shall not only include the assessment results in figures and graphs, but also recommendations and planned actions for closing the performance gaps.

Hence, it is advisable to incorporate an executive summary focusing on the specific improvement potential of each utility that can easily be communicated throughout the organization. This implies a significant amount of work for the project team, but is worth the effort in terms of communicating the assessment results and benefit to the participants.

Depending on the scope of the benchmarking effort (how far the project accompanies the utilities in their improvement plans) improvement actions can be outlined and included as an annex to the individual report.

In most cases, utility reports are confidential. Only the figures from the actual utility are visible. The results of other participants must be reported anonymously. It is the responsibility of the project team to avoid any traceability of individual results while generating the reports.

An outline for a utility individual report is given in chapter 8.6. Annex B, chapter K gives several examples how utility reports can be designed.

8.4.2 Consortium report

The consortium report is often combined with the individual reports. Its aim is to describe the general context of the participating water utilities, their structure and clustering into peer groups of more or less homogeneous performance conditions. The consortium report summarizes the differences between the peer groups, even providing evidence on why it is not recommended to compare utilities of different peer groups (e.g. compact urban networks should not be compared with rural networks in terms of cost efficiency).

Additionally, the consortium report describes which factors generally influence each performance measure (explanatory factors). This facilitates an easier cause-effect analysis for the participants while interpreting their own results.

8.4.3 Public report

The aim of a public report is to achieve a certain degree of transparency. The stakeholders of water services (customers, public supervisory bodies, donators, employees, general public etc.) are informed about the benchmarking effort and its outcomes. Due to the confidentiality of many benchmarking activities, the results are published on an aggregate level. For that very same reason, the public report is first drafted and discussed within the project steering group before is published.

It is recommended that the public report is made freely available to all interested parties, for instance through the Internet, to allow for a wide dissemination of the project.

An outline for a public report is given in chapter 8.6. The full text of some public reports can be downloaded from the Internet (see Annex B, chapter K).

Utility

Once the draft reports reach the utility, the actual project begins. No matter now much time and work have been spent on data acquisition until then, all those efforts will not pay off until they are transformed into actual improvement actions.

Utilities are strongly advised to analyze assessment results in due time. Specifically, utilities should:

Be prepared for the assessment workshop in time
- Spread the draft report internally to the assigned benchmarking team.
- Analyse the draft results internally.
- Get in touch with the project team if anything is not clear.

Document and communicate remarks to the project team
In order to:
- Correct data and suggest amended figures and graphs for the final report.
- Incorporate explanatory facts in the final report.
- Provide drafts of improvement actions or examples of good practices for the workshop.
- Provide feedback on the assessment system itself (definitions, comparability, missing facts etc.)

Have an open mind and avoid subjective bias
- Be ready to accept any kind of results, and not simply reject assessed weaknesses too early. Try hard not to blame everything on context and incomparability.
- Consider performance gaps as an opportunity for improvement.

(Continued)

> **Utility** *(Continued)*
>
> **Keep the assessment results of the individual report as confidential as required**
> - Benchmarking is not a competition or a sports event. It is all about improvement by sharing information among the participating utilities.
> - Utilities should not make ostentation of being the best in class for a certain performance measure, as this might affect other participants.
> - If a utility wishes to make public its own project results (e.g. to customers), a respectable and well-balanced information of its own assessment results should be given, with focus on gaps and the potential for further efforts.

8.5 ASSESSMENT AND BEST PRACTICE WORKSHOP

Closed workshops among the participating water utilities together with the operational project team are the crucial link between the assessment and the improvement phase of a benchmarking exercise. Therefore, workshops should not stop at the analysis of the comparative assessment, but also focus on exchanging experiences and stepping into performance improvement.

The objectives of an assessment workshop are:
- Getting a common view on the general assessment results by presentations of the draft reports from the project team.
- Analysing reasons for good and poor performances and filtering the different influences of explanatory factors on the performance levels.
- Deriving the keys for good practices (e.g. by especific case studies and examples from leading-edge utilities in a certain performance area).
- Drafting action plans for improving performance (or at least some kind of brainstorming of possible improvement actions).
- General exchange of practical experience among the utilities.

Moreover, the workshop should serve as a platform to discuss the assessment procedure in order to improve this phase in future exercises.

While individual written reports are mostly confidential, the environment of a closed workshop facilitates the opportunity to ease these restrictions, provided all participants agree to this (see Figure 8.11). This allows gaining a better insight into the details of the comparative assessment and therefore more information. An adequate way to achieve this atmosphere, is to split participants into small groups, elaborate different aspects and then report results to the rest of workshop participants.

Annex B, chapter K gives examples for workshop agendas.

Figure 8.11 Assessment workshop of a benchmarking exercise (process level)

Utility

Workshops are a core element of a benchmarking exercise as previously pointed out. The following recommendations are provided to participating utilities:

- Always attend the workshop. The people responsible for the assessed performance areas should be involved in the corresponding workshop discussions.
- Do your homework. Prepare and get the most out of the workshop by analysing the draft report in advance.
- Become and active participant. The more you contribute, the more you will get back from your colleagues of other utilities.
- Handle sensitive issues with care (like poor performance results of others). Bring in suggestions and recommendations based on your own experience, but do not lecture others on how they should do things. It is up to them to decide which way to go.

In addition to project workshopws, utility-internal workshops are recommended to internally disseminate the assessment results and to derive actions for improvement. Attendants can be the utility top management,

(Continued)

> **Utility** *(Continued)*
>
> municipal representatives and utility employees of both operational and service units.
>
> The internal workshop can be supported by the project team and could start with a presentation by one of its members. This external and neutral person can also act as a facilitator during the rest of the workshop (interpreting the results and planning of improvement actions).

8.6 FINAL REPORTING ON PERFORMANCE ASSESSMENT

8.6.1 Preparing final reports

Final assessment reports are prepared after comments have been received subsequent to circulation of the draft report. Mostly this follows a common assessment and best practice workshop as described above. Updates may include revised sections of the draft report, reflecting feedback from utilities.

Moreover, additional information from the workshop might be attached, like summaries of best practice presentations, results of lessons learned, data that has been gathered etc.

8.6.2 Disseminating final reports

The individual assessment reports are disseminated to the utilities. They can now serve as the basis for the performance improvement phase.

Depending on the design of the improvement phase within the benchmarking project, the final public report may be disseminated at a later stage, including also elements of the improvement phase.

In connection with completing this product, presentation or communication materials can be prepared. This can serve to aid in writing of articles for journals and publications, for presentations at conferences and seminars, and to publicize and recognize the project and utilities involved.

Following is an example outline of an executive summary, covering the major headings from a final public report:

Confidentiality Statement
1. Executive Summary
    1.1   An Overview of the Benchmarking Study
    1.2   Purpose of the Project-Wide Report

1.3   Environmental Forces Facing the Water Sector Participating Utilities
1.4   Participant Group Business Drivers
1.5   High Level Benchmarking Results
1.6   Leading Practice Themes
1.7   Major Improvement Initiatives
1.8   Historical Trends in the Benchmarking Results
1.9   Best Practice Workshop and Learning
1.10 Concluding Remarks/Lessons Learned/Planned Future Actions

Utility

Disseminating results within utility and stakeholders
As described above, and in order to derive maximum benefit, utilities need to use the assessment results to evaluate strengths, weaknesses and improvement opportunities. Findings need to be examined, and next steps identified. Based on individual utility performance and characteristics, utilities can be paired to optimize improvements.

Individual presentations and internal workshops at utilities
Individual utilities should review the assessment results internally; together with utility leadership teams or other involved groups. These meetings may be organized also together with consultants or coordinators, in collaboration with the utility coordinator and team. The agenda can include:
- Group findings, trends, and recommendations.
- Comparisons across and between utilities.
- Best practices and lowest ranking practices.
- Areas of strengths and weaknesses.
- Conclusions that can be drawn for the own utility.

Often utilities promptly jump into the improvement phase and prepare a roadmap, or action plan for the recommended improvements (see chapter 9). However, this should be done subsequently to discussing the observations and finding performance conclusions.

Following is a sample outline of a typical utility report, including already elements of the improvement phase in the last chapter.
Confidentiality Statement
1. Utility Report – Executive Summary
    1.1 Project Purpose and Scope
    1.2 Utility Comparison to Overall Benchmarked Group
    1.3 Business Drivers
    1.4 Utility Strengths
    1.5 Next Steps

(Continued)

Utility *(Continued)*

2. Benchmarking Participant Group
   2.1 Participant Group
3. Comparative Benchmarkign Results
   3.1 Comparison to Overall Benchmarked Group
   3.2 Regional Comparison
   3.3 Peer Utility Comparison
4. Business Drivers Analysis
   4.1 Utility Business Drivers
   4.2 Comparison of Business Drivers by Region
5. Utility Strengths
6. Utility Key Opportunities and Process Gaps
7. Utility Initiatives for Improvement

Chapter 9

Improvement actions

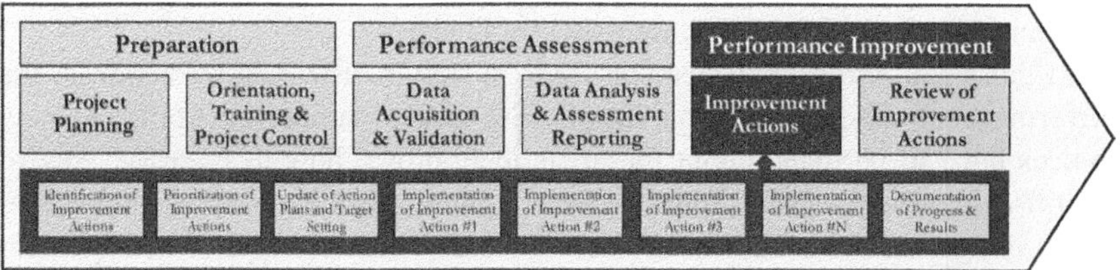

Once the analysis stage has ended, the performance assessment phase is concluded. As a result, participating utilities have learnt about their performance gaps and the challenge is identifying what to improve, how to improve it, and to establish the necessary action plans, including prioritization.

9.1 IDENTIFICATION AND PRIORITIZATION OF IMPROVEMENT ACTIONS

In previous steps, the performance of the participating utilities was assessed. All the different areas under study are brought and considered together. Results were reported to the participating utilities, usually in a written form, but the information might also be made available via a digital report or a personalised Internet webpage. This way, utilities were able to view the variables they submitted and the performance indicators that are calculated out of these. Furthermore they were able to identify their own position against the other participants by means of tables, graphs, or text.

For improved comparisons, grouping was probably applied to take into account the most commonly known explanatory factors for performance differences like the type of water resources or the scale of an area under study. These deeper analyses are often required in order to draw distinct performance conclusions from the performance observations.

With this information, utilities know in what area they perform well compared to the other participants, in what area they perform relatively poor and how large performance gaps really are.

Sometimes benchmarking exercises end right here, with glossy reports for external communication. However, at this point the benchmarking process is just mid-way and to get real added value out of the exercise, it is essential to go on with the next stage, the performance improvement part.

The performance improvement stage is not just the most essential part of the exercise. It is also the most challenging part of all. Getting real performance improvement requires a lot of efforts from the participants. Efforts, in further investigating why performance gaps occur, looking for improvement actions, sharing knowledge and experiences with benchmarking colleagues, having an open mind for new ideas and being ready for changes.

At this point, utility management needs to step in. The data collection in the performance assessment stage is usually the task of supporting staff, but now choices have to be made about the areas to improve, how to improve, how to prioritise between possible improvement actions, decide on action plans and assign budgets to start the improvement actions. These are typical management tasks and general management, operational management and experts on specific processes need to be involved in this process.

9.1.1 Identification of improvement actions

Once the performance gaps are known, the process of identifying improvement actions can start. As a result, utilities should prepare a list of prioritised improvement actions, ready for decision making and implementation.

9.1.1.1 Type of improvement actions

Before discussing the ways to identify improvement actions, one should understand the different types. First type of action is **optimising the existing technologies and working methods**, for instance by adapting the treatment process settings or changing the frequency of billing. In this category the "quick wins" can usually be found: actions without the need for large investments, with a short implementation time and quick results.

A second type of improvement actions is the **application of new technologies, innovations, good/best practices of working methods or**

organisational solutions. Because the implications of implementing this type of improvement action are usually much larger than optimising the existing situation, more research is required and more complex decision making at a higher management level.

Obviously, each of these types of actions has its own advantages and disadvantages, like improvement potential, using the available knowledge of utility professionals, time for implementation, etc.

9.1.1.2 Sources of information

A next question is where to find information about potential improvement actions.

The first source of information for the utility is the utility itself. Due to their daily involvement, utility staff have a lot of knowledge about the presently applied technologies, processes and workflows. Usually, they are very well aware of the actual performance, including strong and weak points. In addition they are a good source for ideas for improving the existing situation. Therefore, utility staff is an important, first source of information.

Externally, there are numerous sources of information. These include, but are not limited to:

- Colleague utilities.
- Professional magazines.
- Newsletters.
- Internet.
- Professional associations.
- Water utility associations.
- Conferences.
- Universities.
- Manufacturers.
- Consultants.
- Other industries with comparable processes.

Water utility associations are a particularly interesting source of information because they know the business the utilities are in, are aware of common issues and have a large network around them with partners that might be able to provide desired information ("matchmaking").

9.1.1.3 Information collection methods

In order to gather information about potentially interesting new technologies, innovations, better working methods and good/best practices a benchmarking workshop can be organized, with the objective to learn from each other. Best performers present their success stories and enable the other participants to translate this to the own situation and identify ways for improvement. In practice this way of exchanging good/best practices has proven to be a good instrument for business improvement.

Of course, organising a benchmarking workshop is not the only way to find information for improvement actions. As described previously, there are many more sources of information than the network of participating utilities in a benchmarking program. Sources of information that might lead you to even more innovative technologies or superior practices. The information from these sources could for instance be disclosed by organising:

- Internal interviews are a way of disclosing the internally available knowledge. Advantage of this option is that it contributes to involving utility staff in generating ideas for improvement and that it helps to create support for changes.
- Literature search aims at finding information about an area under study, getting an overview of the status quo and identifying further sources of information. By using the strong capabilities of the internet, one can save a lot of time and (travelling) costs.
- Company visits help to understand why a certain utility/process works better than the other one. It can also help to generate new ideas for the own situation.
- Site visits show how technologies/processes work in practice and contribute to assess the feasibility of applying the technology and processes in the own situation.
- Calling in assistance from consultants might be helpful when the own organisation has little experience in an area under study or lack of staff. Depending on their track record, consultants might be able to advise on ways for improvement and show cases of improvement actions they were involved in.
- Additional research might be necessary when there is no or little information available in a certain area that needs improvement. This might be organised at individual level, or by joining programs of joint research initiatives, professional associations or water associations. For instance, the IWA Specialist Groups and the AWWARF activities offer many opportunities at research level.

9.1.2 Prioritizing improvement actions

The identification of potential improvement actions will result in a list of possible actions. But how to choose between them, how to prioritise? Indeed, each of the proposed improvement actions will contribute in a different way to the utility objectives. But usually there will be far too many potential actions to implement at the same time, considering the available resources and budget. So, the challenge for utilities is to prioritise the actions properly before decision making and implementation.

Table 9.1 Prioritizing improvement actions

Step 1: determine assessment criteria and weight factors. Relate potential improvement actions to assessment criteria

Criterion	Weight factor	Area 1 water quality			Area 2 reliability		Area 3 service quality
		action 1	action 2	action 3	action 1	action 2	action 1
		replace chlorine by UV	introduce softening	replace led service connections	extra transmission main	extra distribution pump	quarterly instead of monthly bill
Compliance with legal requirements	10	+	+	+	+	+	
Leakage reduction	6			+	+		
Efficiency	4						+
Customer water quality perception	8	+	+	+			

Step 2: determine benefits and cost for proposed actions

Benefits (scale 1 - 10)	Area 1 water quality			Area 2 reliability		Area 3 service quality
	action 1	action 2	action 3	action 1	action 2	action 1
Criterion	replace chlorine by UV	introduce softening	replace led service connections	extra transmission main	extra distribution pump	quarterly instead of monthly bill
Compliance with legal requirements	7	5	4	6	8	
Leakage reduction			3	2		
Efficiency						6
Customer water quality perception	9	5	3			

Costs x 1.000.000 (revenues: negative costs)

Net present value	7,5	3,0	5,0	5,0	1,5	-4,0

Step 3: calculate total, weighted benefits for proposed actions

Criterion	Area 1 water quality			Area 2 reliability		Area 3 service quality
	action 1	action 2	action 3	action 1	action 2	action 1
	replace chlorine by UV	introduce softening	replace led service connections	extra transmission main	extra distribution pump	quarterly instead of monthly bill
Compliance with legal requirements	70,0	50,0	40,0	60,0	80,0	
Leakage reduction			18,0	12,0		
Efficiency						24,0
Customer water quality perception	72,0	40,0	24,0			
Total weighted benefits	142,0	90,0	82,0	72,0	80,0	24,0

Step 4: calculate weighted benefits/costs ratios. Rank proposed actions

	Area 1 water quality			Area 2 reliability		Area 3 service quality
	action 1	action 2	action 3	action 1	action 2	action 1
	replace chlorine by UV	introduce softening	replace led service connections	extra transmission main	extra distribution pump	quarterly instead of monthly bill
Total weighted benefits/costs ratio	18,9	30,0	16,4	14,4	53,3	-6,0
Priority	4	3	5	6	2	1

A simple way of comparing different options is by determining cost/benefit ratios. In this case multi-criteria analysis should be applied because the proposed actions will usually contribute to multiple utility objectives. For this, first of all assessment criteria need to be determined. Usually, these include complying with legal requirements and the strategic objectives of the utility.

To assess the benefits of a proposed action, utility management needs to estimate to what extent the action contributes to the assessment criteria. By assigning weight factors to the different criteria, one can determine the total, weighted benefits for each action.

With respect to costs of proposed improvement actions, it is important to take into account the necessary investment, the operating costs during the lifetime and the removal costs at the end of the lifetime ("total costs of ownership"). By determining the net present values, actions with different lifetimes can properly be compared.

In the end, by calculating the weighted cost/benefit ratios one can rank the proposed actions and assess priorities.

The following fictitious example for a water supply utility illustrates how alternative improvement actions can be prioritised:

This is just an example; the evaluation can of course be further extended and tailor made by introducing assessment criteria like project duration ("quick wins") project complexity, etcetera.

An example for an action plan is given in Annex B, chapter L.

9.2 IMPLEMENTATION OF IMPROVEMENT INITIATIVES

9.2.1 Establishing a plan to implement the improvements

Utility

From a utility perspective, once the specific improvements have been identified, scoped, scheduled, connected to target values and resourced, and committed to by the utility, the work effort needs to be managed as a project in order to assure effective implementation. There are various forms of management and tools available to assist in this process, but it is critical that it be organized, managed and reported as a project or series of projects. A project manager or coordinator should be established within the utility, and clear accountability is necessary.

(Continued)

Utility *(Continued)*

Regular reporting against milestones, budget and expected results will be required. Project teams can be established, with specific charters, deliverables, roles and responsibilities, reporting requirements and performance expectations for the team as a whole, as well as for individual members. The most robust teams may identify critical success factors, a communication plan, key stakeholders, and employee engagement process, and then set a process in motion to measure progress against specific performance measures. The composition of the teams should include subject matter experts, process owners and hopefully include broad functional and hierarchical representation. Often, a Steering Committee is formed within the utility to guide the effort, or there is a regular reporting protocol or relationship to utility leadership.

**Verification & update of action plans, targets &
new benchmarking needs**
Improvement initiatives are best suited or accomplished by utilities that have an annual or regular process for updating work plans or activities, budgets, capital programs and performance targets. Benchmarking action plans will need to be updated, based on their progress and any resulting impacts. This could mean adjustments to staffing, costs, schedules or scopes. Such modifications will be needed not only to assure currency, but also to adjust priorities and align with priorities, strategic plans, budgets and capital requirements.

Documenting results
Regular reporting and documentation of progress and results should be built into the improvement plans. This may be best achieved by creating regular, standard improvement initiative reports. This assures accountability, clarity of purpose, increases efficiency of the effort, and creates greater ability to adapt to necessary changes as the improvements are implemented.

9.2.2 Example of best practice implementation

Costs of distribution O&M in the Dutch water sector

Since 1997, the average costs for operating & maintaining water distribution networks in the Netherlands have decreased by 32% (see Figure 9.1). Corrected for inflation, costs even decreased by 45%. Over time, gaps between best- and poorest performers have decreased clearly.

This efficiency improvement is a direct result of Vewin's national benchmarking program, in which all Dutch utilities participate and intensively share best practices.

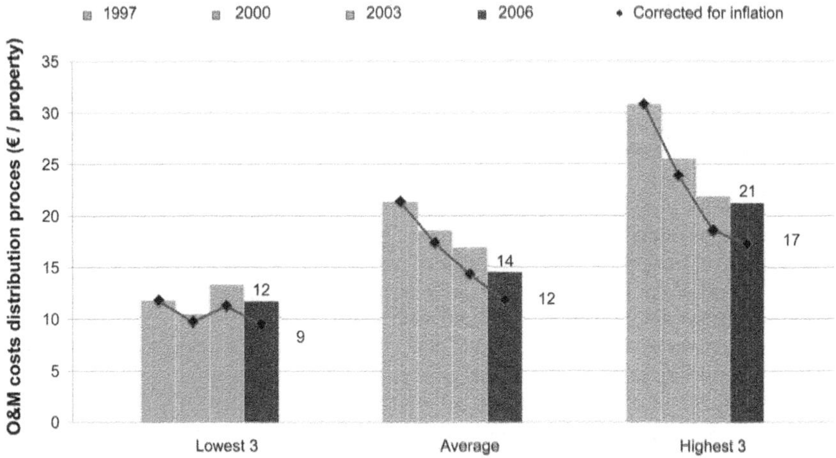

Figure 9.1 Distribution O&M costs Dutch water utilities 1997–2006

In the 1990's, the former water utility Waterleiding Maatschappij Overijssel redesigned the traditional way of operating and managing a water distribution network. By introducing a new work process and developing supportive field automation (planning software on laptops and pda's for maintenance staff) the efficiency of the water distribution operation & maintenance process increased significantly.

Today, maintenance staff start their work day at home. They login to the utility's office network, look at their planning for the day and hit the road for their first job. An online connection to the company's Geographical Information System allows reviewing the actual situation of the network (Figure 9.2). After finishing the job, relevant data like the time spent on the job, materials used and network data updates are uploaded to the office systems and staff can head for the next job.

Company vans are equipped with a (limited) stock of necessary supplies. When the stock runs out of supplies or if additional support is needed (maintenance employees basically work alone), the office is contacted and new supplies or additional staff will be directed to the job site. This way, maintenance staff is more capable to plan the daily work themselves and organise it in a more efficient way. Because company vans are equipped with GPS, the supervisor knows exactly where maintenance staff is located and can easily redirect them in case of network failures.

Figure 9.2 Mobile office. Source: Vitens water company

Once a week or two-weekly, maintenance staff meets in the office to coordinate. At the same time, the company van stock is recharged.

This practice has now been adopted by almost all Dutch water utilities, and has resulted in a significant decrease in labour intensity and a more efficient asset management.

Chapter 10

Review improvement actions

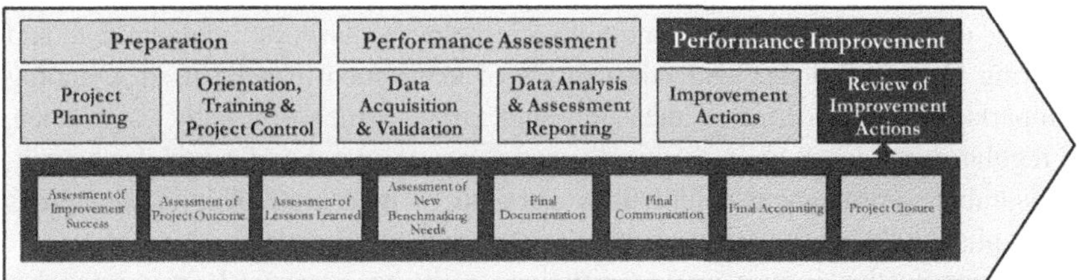

Once improvement actions have been implemented, results have to be evaluated with the objective to check if the previously identified performance gaps are closed. This assessment is the start of an update of the action plan and the definition of new benchmarking needs.

Objectives of this step

- To check at utility level if the previously identified performance gaps are closed.
- To evaluate success of the benchmarking and improvement efforts, including benefits, both at utility and project level.
- To assess new benchmarking needs.

Results of this step

- Determination of performance and improvement gap closures.
- Assessment of project outcome and lessons learned for next benchmarking activities.

- Assessment of the need for new benchmarking activities.
- Project closure.

10.1 REVIEW OF IMPROVEMENT ACTIONS

10.1.1 Assessment at the utility level

The review process starts with the evaluation of each implemented improvement action. Project evaluations usually pay most attention to the question if a project has been executed consistently with the planned scope and budget. However, even more important is to understand the effectiveness of the actions taken. To what extent does the action really address identified problems and contribute to desired levels of performance.

Next, the core question to answer is whether the implemented actions have resulted in closing the earlier identified performance gaps. If not, new strategies and actions should be considered.

The use of performance measures, employee surveys, stakeholder and Steering Group feedback and other forms of assessment can be used. Ongoing comparisons to performance measures and target values originally established, or regular monitoring against critical success factors may be of particular benefit. Measurable improvement in practices can also be measured. This is sometimes accomplished through structured change management processes.

Ultimately the improvement initiatives will be measured in terms that evaluate the effectiveness of new practices, work processes, life cycle and organizational performance, customer benefits, costs and budgets, employee training and acceptance. Some improvements may alter decision processes, make organizational changes, apply new technology, or initiate other practice changes that are important to continually assess and modify as appropriate. Conducting this type of internal analysis is critical in terms of creating a continual improvement environment and culture in the utility.

10.1.2 Timelines

In practice, it is not always possible to measure results within the timeframe of a single benchmarking exercise or cycle. It may take a longer period before reliable information is available to determine the effectiveness of actions. Such an implementation takes time and results have to be monitored and reported over an extended period. This is especially the case of large investment programs such as replacing cast iron pipelines to improve water quality, where final results are not available at the beginning, although development of a plan and initiation of a program could be reported. In other words, the progress of the Plan-Do-Check-Act cycle may extend over multiple benchmarking exercises.

10.1.3 Continuous improvement

At this point, we are at the end of the benchmarking process. However, benchmarking should not be limited to a single action. After identifying performance gaps and implementing actions to close them, utilities must continue to assess and improve. New technologies come to the market, organisations innovate, work processes can be made more efficient; and what was considered as a good or best practice yesterday might already be outdated today. In this rapidly changing world, business improvement should be a continuous focal point for utility management. Therefore, repeated participation in a benchmarking program is recommended. This enables continual monitoring of performance over time. Also, being part of an active, professional benchmarking network makes it much easier to spot new, innovative improvement opportunities.

10.1.4 Comparing benchmarking results across years

An example of continual utility improvement is illustrated below (see Figure 10.1). Benchmarking participants were assessed, between 2004 and 2008, in seven functional areas of asset management, as part of the WSAA asset management project. As can be observed, utility focus in a number of pSEQractice areas resulted in substancial improvement, as validated by third party reviewers.

This improvement varies from utility to utility, but can be attributed overall to the following types of industry initiatives in practice areas of asset management:

10.1.5 Assessment at the project level

At a project level, feedback from participants will be necessary to determine the extent of efforts across utilities in implementing improvements. One particular method of successfully supporting this is through a structured and facilitated workshop. Collective improvements or progress can be measured against several criteria such as schedule, budget, achievement of service levels, quality of the improvement, as well as impact and performance results.

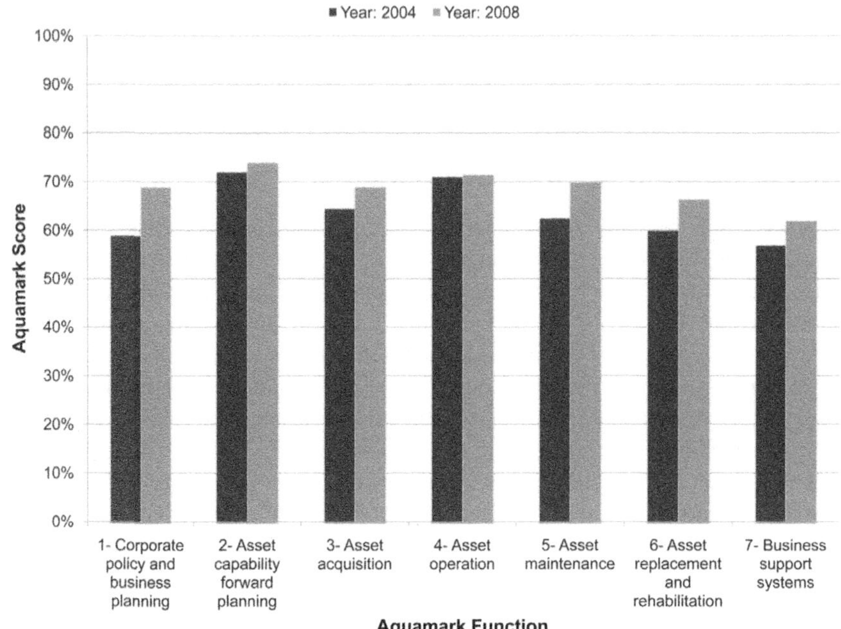

Figure 10.1 2004 to 2008 Performance comparison (WSAA Asset Management Project)

Table 10.1 Comparison of 2004 improvement initiatives to 2008 results

Improvement opportunity	Level of improvement
Configuration management Quality management	Around 10%
Review and improvement Triple bottom line Concept and detailed design Procurement best value Operational work practices Maintenance strategies	Around 5%
Risk and replacement options Level of service Operational strategy development Asset knowledge	Little change

Documentation can be completed by utilities, submitted in advance of the workshop, synthesized by a third party coordinator, and then reported and validated and reviewed at the workshop.

An enhancement to this approach is to incorporate additional training on priority best practices or other challenge areas that were identified during the earlier benchmarking project. The third party coordinator would record and then document and publish the proceedings, including a report which would supplement the original benchmarking findings, recommendations, best practices and quantitative information.

Of lesser value, but still a worthwhile effort is to obtain some form of survey information, usually by means of a questionnaire, to update the original benchmarking project information. This could include performance or metrics data that could be analyzed and compared to a baseline by a third party. This survey may also seek input or suggest topics or plans for future benchmarking projects.

Annex B, chapter M serves with an example of a feedback questionnaire.

Finally, a common practice is utility to utility exchanges or visits. These can prove to be educational, are sometimes geographic based to promote greater staff participation, and can explore certain functions, practice areas, or data sharing. On occasion, these utility liaisons may occur at conferences where participants would be convening anyway.

10.2 FINAL EVALUATION AND DOCUMENTATION OF THE PROJECT. LESSONS LEARNT

At the conclusion of the benchmarking process, the exercise itself should be evaluated by the project coordinator and the participants. Part of the evaluation could cover the performance assessment methodology, the project plan, utility benefits, as well as lessons learnt from the project so as to improve the next exercise.

A report can summarize final results and be distributed to all participants. Some level of quality control or auditing of data quality can also be undertaken. This report can be distributed as a final report; or in some instances, it can be an annual report that summarizes progress against plan. Sometimes the plan is also updated. Lessons learned, provided by participants, can be a benefit to each utility as well as other interested parties.

This step may also include preparation of written communication materials, power point presentations and other documentation that will support each utility's internal training and development efforts. It will also serve as a basis for additional water sector information sharing on the topic at conferences, in journals, for research projects, and at training events. At this stage, utilities should already be thinking of and reporting new benchmarking needs, including priority areas to explore, methodological improvements, as well as technology and innovation applications.

10.2.1 Project closure

After filing all relevant project documents and communications by the Project Coordinator, the benchmarking project is closed.

When benchmarking projects are part of the continual process of a benchmarking program, a long term relationship is established between utilities and the program facilitators and maintained around efforts to benchmark, report, share practices and build into improvement processes of utilities. Nevertheless, each benchmarking project has a definite scope and life, so it is important to bring closure to the project, as originally defined, and then move on to the next one.

Annex A

Benchmarking efforts in the water industry

Since the early 1990's, benchmarking has been in constant evolution in the water sector. Most of the early efforts consisted in the comparison of metric figures, but they later evolved into more complex schemes seeking the improvement of performance.

In this chapter we have tried to provide some examples of key benchmarking initiatives in the water industry in those years. The selected projects are not necessarily the most important or relevant, but were chosen to illustrate the evolution of benchmarking efforts in the water industry.

A. OFFICE OF WATER SERVICES (OFWAT)

The regulatory work carried out by OFWAT in England and Wales set out some of the industry basics regarding performance assessment and its comparison. Created in 1989, OFWAT soon made of the yardstick competition one of its main regulatory tools for the newly privatized water industry in the country. As a consequence, metrics were collected and audited and indicators calculated and compared. The need for a transparent and equitable model led to the development of communication techniques (i.e. graphs and tables) as well as important concepts like confidence grading[1].

---

[1]Later adopted (and adapted) by the IWA system of performance indicators, the OFWAT confidence grading system assessed the reliability and accuracy of all data used to calculate their performance indicators.

Table A.1 Overview of some benchmarking efforts in the water industry

Programme name	Country	Programme type	Level of detail	Type of activity	Geographical scope	IWA manuals based
6-Cities Group	Scandinavia	BM	U, F & P	WS & WW	R	no
DANVA	Denmark	BM	U & F	WS	N	no
European Benchmarking Co-operation	Europe	BM	U, F & P	WS & WW	R & I	yes
Germany (several)	Germany	BM	U, F & P	WS & WW	N	yes
NWWBI	Canada	BM	U, F & P	WS & WW	N	no
OEWAV	Austria	BM	U & F	WW	N	no
OVGW	Austria	BM	U & P	WS	N	yes
QualServe	USA	BM	U	WS & WW	N	no
SEAWUN	South-East Asia	BM	U	WS	R	no
Vewin	The Netherlands	BM	U, F & P	WS	N	no
WSAA	Australia	BM	F & P	WS	R & I	no
ADERASA	Latin America	PA	U	WS & WW	R	no
FIWA	Finland	PA	U	WS & WW	N	no
IBNET	World Bank	PA	U	WS & WW	I	no
Norsk Vann	Norway	PA	U	WS & WW	N	no
OFWAT	England & Wales	PA	U & F	WS & WW	N	no
Svensk Vatten	Sweden	PA	U	WS & WW	N	no

PA - Performance assessment; **BM** - Benchmarking (Performance assessment & improvement); **U** - Utility level; **F** - Functions level (core processes); **P** - Process level; **WS** - Water supply; **WW** - Waste water; **N** - National; **R** - Regional; **I** - International

In 1996, seeking additional motivation and competition for the regulated companies, OFWAT started to participate in several international projects (mainly with Australian utilities). These initiatives still continue today, mainly with the participation of other regulated utilities[2].

However, the companies themselves do not take part in these exercises and the projects mainly consist in exporting the OFWAT model and comparing with the data obtained from England and Wales for regulatory purposes.

B. THE WORLD BANK

If the OFWAT is a clear representative of regulators (always interested in performance assessment) the World Bank would be the best example of a funding agency interested in promoting transparency and investment prioritization through the comparison of performance indicators.

In 1999, the Bank published its "benchmarking start-up kit" which comprised a series of documents and a spreadsheet based software to collect data from water utilities worldwide. This kit was the precursor of IBNET an online database (the world's largest according to the website, http://www.ib-net.org, see Figure A.1) for water and sanitation utilities performance data.

While the wealth of data in IBNET is certainly astonishing, the fact that is a voluntary network and the obvious difficulties in assuring data quality limits the usefulness of the system. In any case, both academics and practitiones are likely to find useful information with relevant searches in the system.

C. WATER SERVICES ASSOCIATION OF AUSTRALIA (WSAA) BENCHMARKING

WSAA is the peak body of the Australian urban water industry. Its Members provide water and wastewater services to approximately 15 million people in Australia and New Zealand. Since 2000 WSAA has sponsored a rolling program of process benchmarking exercises, with IWA co-sponsoring the program since 2007.

The programs have been:
- 2000 – Civil Maintenance Practices (13 utility participants).
- 2001 – Mechanical and Electrical Maintenance Practices (14 utility participants).

---

[2] In 2008, Scotland, Northern Ireland, USA, Portugal, The Netherlands, Canada and Australia (http://www.ofwat.gov.uk/regulating/reporting).

- 2002 – Customer Service (12 utility participants, plus international comparisons).
- 2003 – Shared Services (10 participants, plus limited UK data).
- 2004 – Asset Management (23 utility participants).
- 2005 – Civil Maintenance Practices (19 utility participants).
- 2006 – Mechanical and Electrical Maintenance Practices (17 utility participants).
- 2007 – Customer Services (15 utility participants).
- 2008 – Asset Management (42 utility participants).

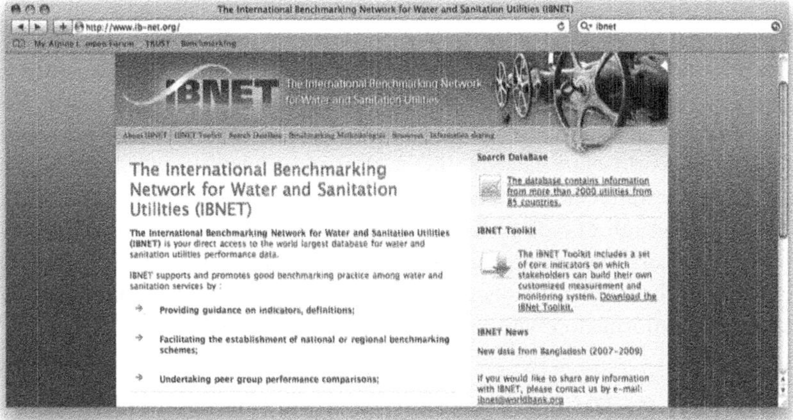

Figure A.1 IBNET website

WSAA developed the "Aquamark" model as an industry-based framework for asset management process benchmarking (see Figure A.2) as a result of earlier exercises identifying its need. The core purposes were to:

- Provide meaningful water industry comparisons in asset management practice.
- Promote retention of industry learning and intellectual property.

Implementation services are encouraged beyond the scope of the benchmarking projects to facilitate implementation and engender ownership by the participating utilities in the project outcomes to gain the benefit from their investment.

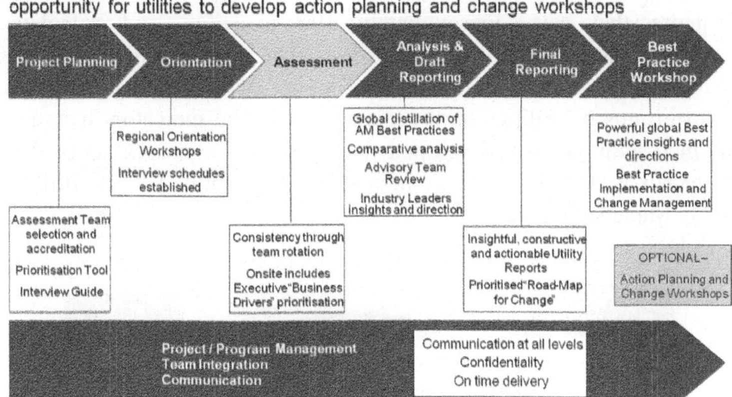

Figure A.2 The WSAA benchmarking approach

D. AMERICAN WATER WORKS ASSOCIATION (AWWA) AND WATER ENVIRONMENT FEDERATION (WEF) – QUALSERVE

The American Water Works Association (AWWA) administers a QualServe program (see Figure A.3) that was created in the mid-1990's, providing a means in North America for water and wastewater utilities to continually improve using a Plan-Do-Check-Act framework. It currently offers a well-established toolbox of a self-assessment, peer review, and benchmarking for utilities. The self-assessment consists of an employee survey to gauge their opinions and to build buy-in and support for improvements. It asks questions that cover five business systems that are typical to water and wastewater utilities in North American, including: Leadership and Organizational Development; Business Operations; Customer Relations; Wastewater Operations; and Water Operations. The Self-Assessment is a customizable electronic survey and report instrument that allows a utility to perform an internal examination of its practices compared to "best practices" as defined by 430 statements. The Peer Review is a third-party review of the utility's performance by experienced professionals and managers from other water/wastewater utilities. The peers are specialists in one or more of the business areas and categories being assessed. Since its inception, approximately 135 utilities have conducted self-assessments and over 100 have completed peer reviews.

In addition, this program includes a metric benchmarking element. On an annual or biannual basis, a survey is circulated and completed by over 300 water/wastewater utilities. The current Benchmarking Survey is an electronic questionnaire that asks approximately 190 questions and calculates 34 performance indicators from these questions. The aggregate data is published; and following the survey of utilities, Data Sharing Benchmarking Workshops are offered to groups of utilities on a regional basis. The indicators have recently been aligned with a newly developed management framework adopted by the leading North American water sector associations, entitled "Ten Attributes of Effectively Managed Utilities."

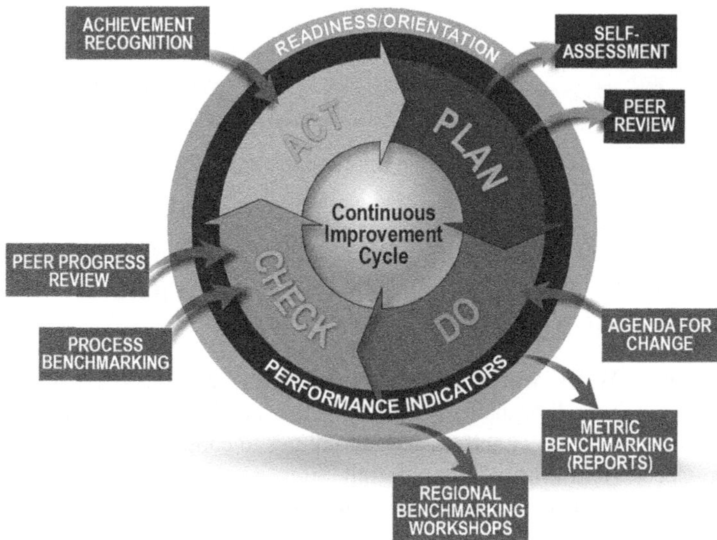

Figure A.3 Overview of QualServe Program

The Water Research Foundation (WRF) conducted two recent projects designed to enhance and increase the value of the QualServe program: Enhancement of Water Utility Self-Assessment Tools to Improve Utility Operations (2006); and Enhancement of QualServe Tools to Improve Utility Operations (2009). A new research and benchmarking project is now being conducted, using the framework outlined in this manual and linked to the 10 Attributes, with the intention of identifying and updating the leading practices associated with these attributes, assessing performance of utilities, and identifying utility improvement plans. This project is entitled, "Performance Benchmarking for Effectively Managed Water Utilities" (2010).

E. AUSTRIAN BENCHMARKING

Water services in Austria are often very small-structured. Besides some large city suppliers, thousands of small water suppliers and decentralised sewage treatment utilities exist due to the good conditions in water quantity and qualityAlong with the issues of privatization and liberalization, the debate on the efficiency of small utilities has been present for a while.

In 1999, the Austrian Association for Water and Waste Management (OEWAV) launched the Austrian wastewater benchmarking program in a pilot project with 78 utilities. Since then, five project runs have been carried out for both sewerage and treatment involving more than 100 wastewater utilities. The objective is to find cost saving potentials on process level and allow utilities to exchange improvement experiences in workshops.

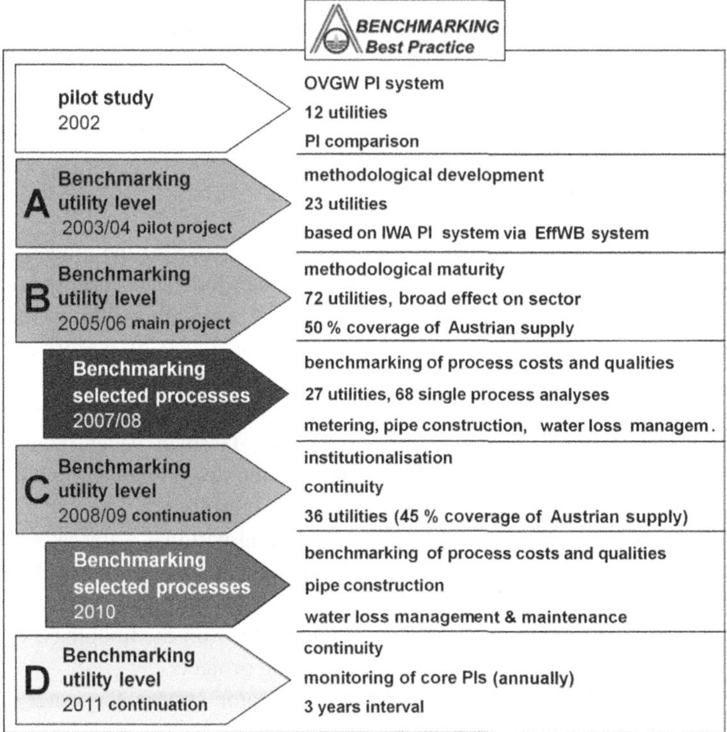

Figure A.4 History and further strategy of the OVGW water supply benchmarking program

After a long tradition in water statistics, the Austrian Association for Gas and Water (OVGW) started its water supply benchmarking activities in 2002. Figure A.4 shows the development of the program designed as a long-term strategy from the beginning.

In 2009, 100 utilities had undergone three full benchmarking cycles at utility level, covering more than 50% of the Austrian water supply. Furthermore, additional benchmarking was carried out on six selected processes, leading into 68 single process analyses for 27 water suppliers. Finally, cross-border comparisons were undertaken with Bavaria, Hungary and Slovenia.

The participation in the Austrian water and wastewater benchmarking programs is voluntary and confidential. Public reports on aggregated levels are published and serve for transparency to the stakeholders.

The Austrian programs are quite unique in the fact that are predominantly executed by university institutes. The reasons quoted by the commissioning associations are their neutral public position, their data management skills and their drive for further developing the benchmarking systems.

Austrian experiences show that small utilities are able to achieve good performance levels within the limitation of their systems, and they can certainly engage in benchmarking efforts and benefit from their outcomes.

F. GERMAN BENCHMARKING

The German water industry's benchmarking concept is part of the modernization strategy for the regulatory framework of the German federal government. It was developed and promoted by the water sector itself in consultation with the political partners. Main objectives were i) to increase transparency of performance (in terms of reliability, water quality and safety, customer service, sustainability) and costs in the water supply and wastewater services and ii) to optimize processes and open up potentials for improvement. For the German water industry the concept of successful benchmarking was based on the two prerequisites: voluntary participation and confidentiality of utilities' data and results.

A conceptual framework provided by the leading national water associations maintained a variety of performance assessment initiatives offered to the individual utilities (around 6,400 water suppliers and 6,900 wastewater utilities). This framework also included guidelines and manuals on methodology, code of practice and quality management of benchmarking projects.

In drinking water, the IWA performance indicator system for water supply services has been widely applied in nation-wide performance assessment studies on the federal states' level. The majority of these projects does not explicitly challenge performance improvement, but significant progress has been made on the scope of individual utilities. Optimization and efficiency gains in water supply are predominantly pushed by a variety of specialized process benchmarking trials, covering the whole range from water catchment, water treatment, meter reading and exchange, etc. (cf. Profile of the German Water Industry, 2008 for detailed project listings).

For performance assessment and benchmarking projects of the wastewater industry, the DWA system of key performance indicators provides as well a common basis for the projects compatible with the international IWA performance indicator system. Similar to water supply, process benchmarking has its dominant role in identifying improvement potentials in wastewater collection and treatment processes, widely accepted by the utilities with significant scope and specialized focus throughout the years.

There has been a tremendous methodological development and progress in supporting tools and communication of benchmarking results over the last five years. Web-based data acquisition and quality control tools offer easy access to data and ensure high data quality.

A shortcoming of the existing institutional scheme of performance assessment and benchmarking is, that the less motivated utilities are still reluctant in joining the initiatives – and very likely, among those will be significant improvement potential. Starting off from 2008, the current system of voluntary performance assessment and benchmarking has become under pressure from rigorous debates on price control schemes by the Federal Trade Commissions, along with an intense discussion on regulatory schemes for the German water supply industry.

Profile of the German Water Industry (2008): www.dvgw.de/fileadmin/dvgw/wasser/organisation/branchenbild2008_en.pdf

G. DUTCH BENCHMARKING PROGRAM

The Netherlands have a long tradition on national water statistics and benchmarking. In the 1980's, the association of larger regional water utilities COCLUWA started developing a set of comparative performance indicators. In 1991 first comparisons were made, driven by the need of utility managers of better management information. In 1997 the Dutch drinking water sector started a benchmarking program at a national scale. Although voluntary, the program was initiated under the pressure of discussions on liberalisation and privatisation of public services all over Europe.

The benchmarking program, co-ordinated by Vewin, offers a wide view on utilities' performance. Water quality, service quality, environmental impacts and finance & efficiency are thoroughly analysed every three years. The results of these exercises are made available to the public and to stakeholders to increase transparency.

Several best-in-class workshops have been organised to exchange knowledge and best practices. Additionally, and to monitor results, every year a financial assessment is made at process level with a closed model that is fully linked to the annual account.

Over a period of 11 years since the start of the program, Vewin reports an efficiency improvement of the water companies of 26 percent (as calculated with a DEA-model). Although this performance improvement cannot be attributed exclusively to the benchmarking exercises, and also derives from mergers and restructuring programs Figure A.5 illustrates the potential benefits of a longer lasting benchmarking program with a continuous management focus on performance gaps and improvements.

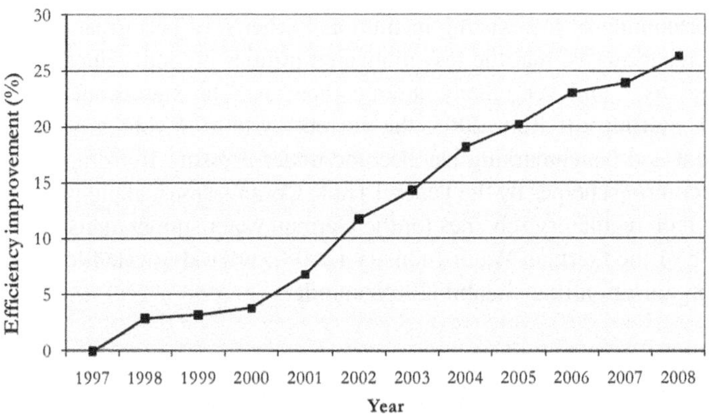

Figure A.5 Efficiency improvement in the Dutch water utilities

The recently renewed Dutch Drinking Water Act reflects the political decision to keep the drinking water sector public and promote efficiency by introducing mandatory benchmarking. This mandatory program will largely be based on the present voluntary benchmarking scheme, which underlines the value of a having a strong sector program.

H. 6-CITIES GROUP

The Scandinavian cities of Copenhagen, Oslo, Helsinki, Stockholm, Gothenburg and Malmo have a long tradition of co-operation in development projects.

In 1995 the group started a project with the aim of developing performance indicators (PIs), in order to facilitate comparisons between the cities.

The group developed PIs within the following areas: customer satisfaction, quality, availability, environment, organisation/personnel and economy. These PIs, compiled for the six cities, provided a basis for a performance comparison between the cities and in-depth discussion as to why some cities had exceptionally high or low PI values. The conclusions from these discussions dinspired new approaches in the ongoing internal change processes in the participating utilities – reinforcing driving forces for quality and efficiency – and constituted a step in the development of a benchmarking process.

This project soon became a reference in Europe. Among other "firsts", a comprehensive software package was developed to not only collect data from participants, but also manage the information and aid in the reporting.

More recently, the group is focussing on assessing overall performance and asset management.

I. SCANDINAVIAN PROGRAM

The Scandinavian countries' benchmarking history dates to the second half of the 1990's. In this case, the national water associations took the lead in organising the programs. DANVA, the Danish water association, has the most extensive program of the four countries, which started as a typical metric performance assessment program. Today, the program is developing into a benchmarking one with the aim to promote performance improvement. Recent changes in the Danish Drinking Water Act will result in mandatory benchmarking for the industry.

Svenskt Vatten from Sweden applies a similar program as in Denmark, facilitated by the same consultant.

The benchmarking programs from Norsk Vann of Norway and FIWA of Finland are quite straight forward performance assessment schemes.

The Scandinavian programs are not country-wide extended. Typical coverage of these benchmarking programs is 40% of the population. Next to several larger and medium scale (municipal) water utilities, there is a large number of small utilities and even smaller private "co-operatives" that cannot be reached easily and have not taken part in the intiatives so far.

J. EUROPEAN BENCHMARKING

In 2004 the European Benchmarking Co-operation was initiated by the Dutch and Scandinavian national water associations and several utilities of the 6-Cities Group.

Objective of this IWA-supported initiative is to allow European[3] water utilities to improve their business processes by offering an international benchmarking program and providing a network to exchange knowledge and best practices.

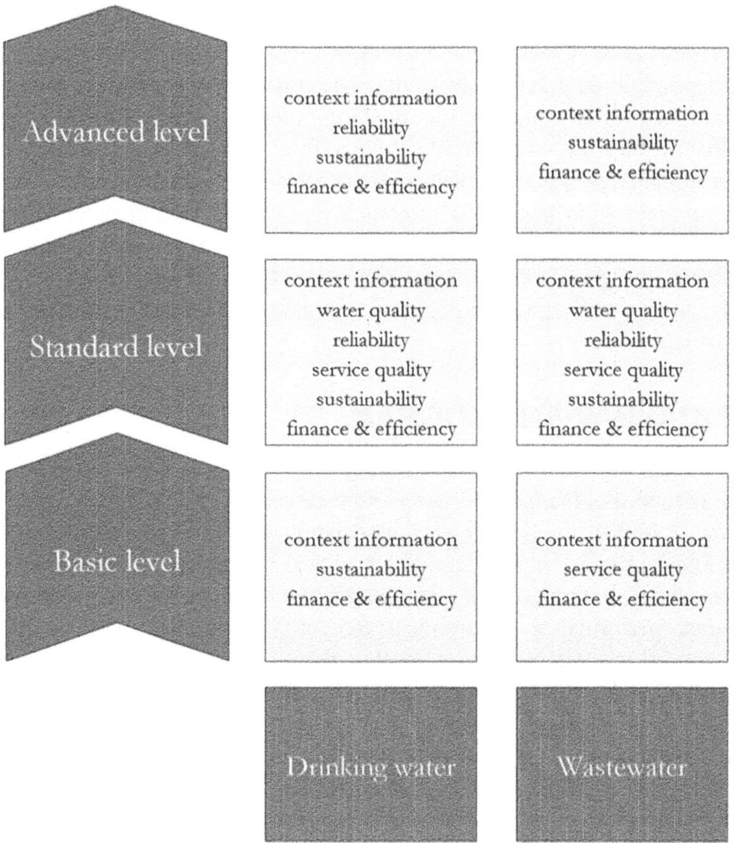

Figure A.6 EBC's levels of participation

Main reason to start this initiative was the "law of diminishing returns". In the Netherlands as well as in some Scandinavian countries, utilities were already benchmarking for quite a long time. Participants in the respective national programs compare their performances, exchange information about technologies

---

[3] Although the main focus of the initiative are European utilities, the co-operation has been recently expanded to utilities from other continents.

and innovations and discuss their business processes. As a result, performance differences between the participants decreases. However, after several years of intensive benchmarking with the same group, utilities cannot learn that much from each other anymore and especially in the Netherlands the need for new impulses arose. To facilitate this, the national associations of the Netherlands and the Scandinavian countries jointly decided to establish a cross-border benchmarking program. Additional to the existing national benchmarking schemes, this program offers utilities the opportunity to connect to a large international network, find new peers and identify new best practices.

The participants in EBC's benchmarking program can choose themselves at what level of detail they wish to participate: basic, standard or advanced.

The advanced level provides utilities with the most detailed insight in their processes and performance. However, reality in Europe is that in most cases water services are provided by rather small scale utilities which are not always able to provide detailed performance information. By offering different benchmarking levels (see Figure A.6) EBC especially encourages smaller utilities to join the network, learn and benefit from it and move forward.

Data is collected via the program website www.waterbenchmark.org. After reporting, participants are invited for a workshop to discuss the assessment results, learn from best practices and prepare for improvement actions.

Since 2007, EBC annually organises international benchmarking exercises. Started as a North European initiative, the program has developed into a European program covering today some 45 water utilities from 21 different countries (of which 3 outside Europe) representing some 55 million inhabitants.

K. SEAWUN

In 1993 the Asian Development Bank (ADB) published a list of indicators from 38 utilities located in the region. A second edition of the publication took the number of participants to 50. This publication later developed into the SEAWUN initiative.

The five South East Asian countries Indonesia, Laos, Malaysia, Philippines and Vietnam decided in 2001 to collaborate in a regional network, considering that large benefits could derive from sharing experiences on common problems and issues. A primary goal of the ADB supported network SEAWUN is performance improvement through benchmarking.

In 2004 performance data of 47 utilities were collected, compared and reported in a data book for the first time. This was followed by a second assessment with almost 90 utilities, reported in 2008. Today the focus of the network is changing from performance comparison towards continuous improvement.

L. ADERASA

ADERASA is an association of regulators for water and wastewater services in America created in 2001. Its objectives are to promote efficiency and efficacy in those services.

One of its working groups is dedicated to benchmarking and was created and published a manual on performance indicators in Spanish, describing methodology and providing indicators. Later, a regional database was created to collect the indicators and organize future benchmarking projects.

In 2003, and within a collaboration with PPIAF[4], the ADERASA benchmarkign program was created, with the objective of providing the association and its members with the knowledge and the necessary tools to apply benchmarking for regulatory purposes. Eventually, all member countries joined the benchmarking project. Since 2004, the group undertakes yearly activities including the collection of data from the previous year and their analyses, an annual meeting to share experiences and results and the ellaboration of a yearly report (see Figure A.7). All reports are available at the ADERASA website[5], and include the analysis of 100 utilities in 16 countries through 58 performance indicators.

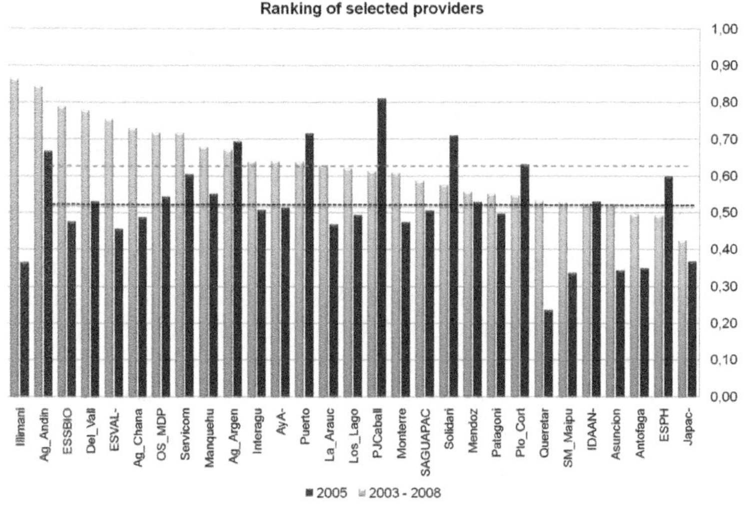

Figure A.7 Utility efficiency ranking in selected ADERASA utilities

---

[4] PPIAF: Public Private Infrastructure Advisory Facility: provides funding to allow governments to explore the public-private partnerships to improve infrastructures.
[5] http://www.aderasa.org/grupos_bench.html

In order to determine global efficiency, ADERASA uses efficiency frontier methods to determine best performers. As additional data become available every year, the analysis can be improved and trends can be set.

M. CANADIAN BENCHMARKING

The Canadian National Water and Wastewater Benchmarking Initiative (NWWBI) was launched in 1997 in response to a need for Canadian municipal water and wastewater utilities to measure, track and report on their utility performance. While fundamentally a high level performance comparison process, it has developed into a network and information base for Canada's most progressive municipal utilities.

The goal of this project is to assist water, wastewater and storm water utilities to continually improve their performance and efficiency. This is accomplished through the implementation of ongoing improvement measures and best management practices.

The Partnership now includes 45 of Canada's leading municipalities and regional districts, representing over 60% of the Canadian population. Participation in this initiative has grown to the point where this initiative serves as the national standard for water and wastewater utility benchmarking in Canada.

The Benchmarking Utility Management Model that was developed in the early phases of the benchmarking project has stood the test of time well. The model has proven to be a robust framework to conduct performance measure in a very meaningful and tangible manner. At the heart of the utility management model are seven key utility goals that have been accepted within the benchmarking partnership:

- Provide reliable service and infrastructure.
- Ensure adequate capacity.
- Meet service requirements with economic efficiency.
- Protect public health and safety.
- Provide a safe and productive workplace.
- Have satisfied and informed customers.
- Protect the environment.
- The utility management model (see Figure A.8) and the benchmarking methodology developed provide infrastructure managers with the means to link their goals, performance measures, strategies and performance monitoring/reporting. In turn, this approach provides managers with the basis on which to strive for superior performance through the development and documentation of best management practices. This framework makes the NWWBI an educational tool for utility managers for decision making

rather than a report card. It's about improvement, not about "Who's Better than Who".

- Key success factors of the program are the way the data are collected by on-site visits instead of questionnaires, the annual workshop, where representatives from participating utilities meet to review and discuss results and to collaborate on improvement plans, and the good history of quality performance measurement data.
- Now, a number of alternative platforms exist to consider, discuss and adopt a growing range of best practices.

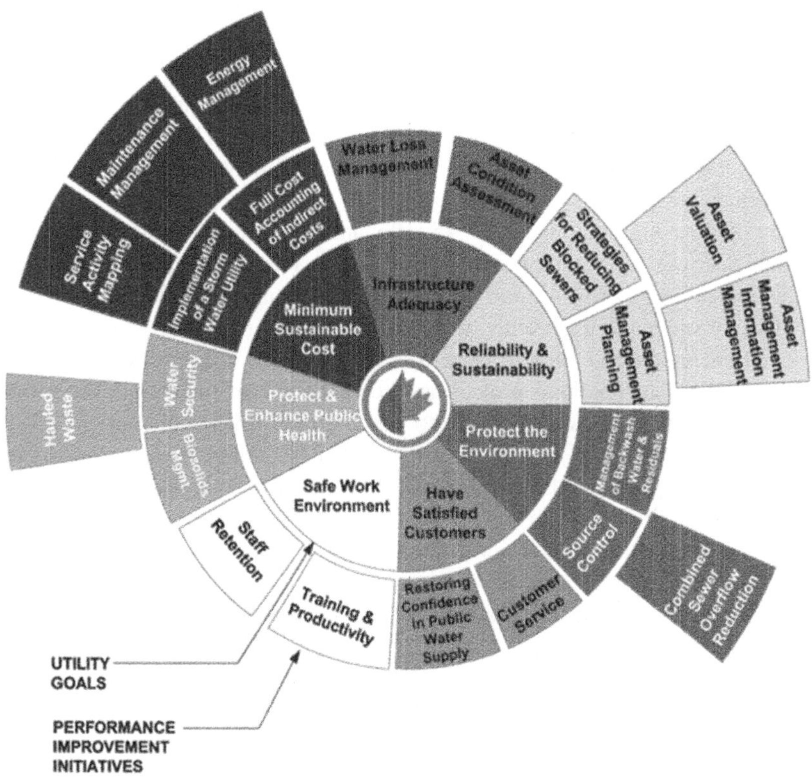

Figure A.8 Canadian benchmarking utility management model

N. PAS PROJECT (INDIA)

The PAS Project is a performance assessment effort in the states of Maharashtra and Gujarat in India that started in 2009 and will span for 5 years. The project intends to develop a statewide performance assessment system, which in total will include over 400 cities and estimated population of 60 milion. These impressive figures could turn PAS into the largest performance indicators project for water services that the world has witnessed.

The project, which is funded by the Bill and Melinda Gates Foundation, is facilitated by CEPT University in partnership with the state governments. The system includes an impressive set of indicators (regarding both the quantity of PIs and the quality of the system) to measure performance regarding coverage, service levels and quality, financial viability, efficiency and equity. The project is very ambitious and intends to gather data from cities ranging from 30.000 to 13 million inhabitants (Mumbai).

Especially remarkable about PAS are the standards to which the project has been designed. Goals have been set, the quality of data is assessed (confidence grading) and in the first round, the data will even be collected by project "investigators" to avoid inconsistencies.

The PAS Project goes well beyond performance assessment, and the organizers support the use of PIs for decision making regariding planning and budgeting. Additionally, the project intends to support capacity building for different stakeholders and identify good practices during visits, in order to disseminate them to all participants and even post them on the project website[6].

Although the results of the project remain to be seen, the effort is certainly based on solid foundations from a technical point of view. At the time of printing this manual, the first data collected from the cities were being analyzed showing already some interesting conclusions.

---

[6]www.pas.org.in

Annex B

Examples of project documents along the benchmarking process

This annex includes documents and examples from real benchmarking efforts to illustrate the benchmarking process described in the manual and provide real examples and further information to the reader. Project organizers will find ideas on how project documents can be structured along the benchmarking process.

The examples should be taken as such, and by no means should be understood as templates or IWA endorsed. For the same reasons, the documents provided may not be complete. The included documents originated in real projects in which the authors were involved, and the organizations and utilities mentioned in the examples provided the necessary permissions regarding property rights and confidentiality of the data.

While the printed annex provides a reference to the document, the actual files can be downloaded from the website of the IWA Specialist Group on Benchmarking and Performance assessment. By providing the files in electronic format, the size of the manual can be optimized (some of the documents would not display adequately in the manual) and files can be updated if needed.

To download the files, please visit the following URL: http://www.iwabenchmarking.com/manual

A. PROJECT PLANS

Basic project planning is done right at the beginning of each benchmarking activity in the preparation phase (see chapter 5 on project planning), and communicated during the recruitment of participants and during the orientation, training & project control phase (see chapter 6).

In this document, the basic framework like project objectives, scope and deliverables is defined. It can be split in more project documents as for the WSAA Asset Management Benchmarking (see files of next chapters).

European Benchmarking Cooperation – Project Plan 2009

Covered key items:
- Project objectives
- Starting points (e.g. scope and assessment model)
- Project deliverables
- Target group
- Project Organization
- Rough schedule
- Participation costs
- Conditions for participation
- Registration

File name: EBC_Project-plan_benchmarking-exercise-IB2009.pdf

B. PROJECT SCOPES

Project scopes:
- must be defined at the beginning of each benchmarking activity in the preparation phase (see chapter 5 on project planning),
- are communicated during the recruitment of participants and
- are sometimes finetuned together with the participants during the orientation, training & project control phase (see chapter 6).

ISO 24500 Guideline Series – Defining Scope and Performance Assessment Model

Covered key items:
- Hierarchical approach of performance assessment
  - Identify components
  - Define objectives
  - Define assessment criteria
  - Define PIs
  - Assess performance vs. objectives

File name: ISO 24510.pdf

European Benchmarking Cooperation – Modular Project Scope

Covered key items:
- 3 levels of detail for the assessment
  - Basic level
  - Standard level
  - Advanced level
- Assessment for
  - Water Supply (single)
  - Wastewater (single)
  - Water & Wastewater (combined)

File name: EBC_Project-plan_benchmarking-exercise-IB2009.pdf (page 3)

WSAA Benchmarking – Scope for Function, Process & Task Level on Asset Management

Covered key items:
- Defining asset management by seven functions
- Breaking down the seven functions to 50 processes, 250 sub-processes & 600 measures
- Defining management and supporting functions

File name: WSAA_Aquamark-User-Manual_Version3.pdf (page 11ff)

Austrian OVGW Benchmarking – Scope for Utility Level

Covered key items:
- Assessment model aligned with five key objectives of performance
  - Supply safety
  - Supply quality
  - Customer service
  - Sustainability
  - Efficiency
- Assessment model based on IWA PI system

File name: OVGW_Project-Scope_Utility-Level.pdf

**Austrian OVGW Benchmarking – Scope for Process
Level on Water Loss Management (1 page)**

Covered key items:
- Defining process by sub-processes
- Defining supporting processes
- Combined assessment model for
  - efficiency (costs, time) &
  - quality (process quality & process results)

File name: OVGW_Project-Scope_Process-Level.pdf

C. PROJECT GOVERNANCE

The project governing structures are set up in the preparation phase (see chapter
5 on project planning) and communicated during the recruitment of participants
and during the orientation, training & project control phase (see chapter 6).

**European Benchmarking Cooperation – Program
Governance**

Covered key items:
- Roles & responsibilities of
  - Steering Committee
  - Programme Committee
  - Expert groups
  - Project Team
  - Regional hubs
  - Participating utilities

File name: EBC_programme governance November 2009.pdf

WSAA Benchmarking – Steering Committee Function

Covered key items:
- Role of Steering Committee
- Members of Steering Committee
- Decision making process
- Role of the Consortium and the decision making process
- Future Dates of Steering Committee meetings and project workshops

File name: WSAA_Function-of-Steering-Committee.pdf

Austrian OVGW Benchmarking – Project Structure

Covered key items:
- Relations of main parties
  - Project governing body OVGW
  - Project team
  - Participating utilities
- Collaboration of parties in BM work group
- Data flow & reporting

File name: OVGW_Project-Structure.pdf

D. MILESTONES AND SCHEDULE

Time schedules and milestones are
- set up in the preparation phase (see chapter 5 on project planning) and
- communicated during the recruitment of participants and during the orientation, training & project control phase (see chapter 6).
- If necessary, they may be updated during the following project phases.

European Benchmarking Cooperation – Time Schedule 2009

Covered key items:
- Project phases
- Time line
- Milestones (e.g. for workshops and deadlines of reports)
- Participants' main responsibilities in each project phase

File name: EBC_Time-schedule_exercise-IB2009.pdf

WSAA Asset Management Benchmarking – Time Schedule

Covered key items:
- Project phases and tasks
- Time line with dates and critical path
- Milestones (e.g. for workshops and deadlines of reports)

File name: WSAA_Time-schedule_Asset-Management-Benchmarking_V4.0.pdf

E. CODE OF CONDUCT AND CONFIDENTIALITY AGREEMENTS

The involved parties need to agree on general terms of reference, especially regarding confidentiality issues. This task shall be achieved before inviting utilites for participation. It can be adapted by specific expectations of participating utilities.

European Benchmarking Cooperation – EFQM European Code of Conduct (6 pages)

Covered key items:
* Introduction to benchmarking
* Principle of preparation
* Principle of contact
* Principle of exchange
* Principle of confidentiality
* Principle of use
* Principle of legality
* Principle of completion
* Principle of understanding and agreement
* Benchmarking with competitors
* Behaviour at face-to-face site visits

File name: EBC_EFQM_benchmarking_code_of_conduct_2007.pdf

OVGW Benchmarking – Confidentiality Agreement (extract from the OVGW participation contract)

Covered key items:
* Scope of the agreement
* Confidential data handling by the project team
* Confidential data handling by the utility
* Exceptions

File name: OVGW_Confidentiality-Agreement_Project-Utility.pdf

F. RECRUITMENT OF PARTICIPANTS

Utilities can be invited to participate in a benchmarking project, once the project milestone "project approval" has been achieved. Different communication channels can be used (see chapter 5).

European Benchmarking Cooperation – EBC Program Leaflet

Covered key items:
- The initiative
- Benchmarking as a structured, continuous process on PDCA basis (plan-do-check-act)
- Modular assessment model
- Yearly benchmarking projects
- Example of PI results
- Why participate?
- Contact

File name: EBC_Flyer_benchmarking-programme.pdf

European Benchmarking Cooperation – Invitation Letter

Covered key items:
- Invitation to a concrete project (e.g. IB2009 exercise)
- Why participate?
- Key project outputs
- Project overview
- Registration until ... with attached registration form

File name: EBC_Invitation-to-register_exercise-IB2009.pdf

G. PROJECT COMMUNICATION

Plans for external and project-internal communication are set up in the preparation phase (see chapter 5 on project planning).

Internal communication both on project and utility-inernal level is finetuned during the orientation, training & project control phase (see chapter 6).

Communicating the progress of the benchmarking effort is useful allthrough the benchmarking process, e.g. by newsletters.

WSAA Benchmarking – Project News Update Dec 2007

Covered key items:
- Project-internal communication of key project info
  - Confirmed participants
  - Key dates
  - Preparation for the project
  - Invoicing
  - Participation agreement

File name: WSAA_Project-News-1_Dec2007_AM-Project.pdf

WSAA Benchmarking – Project News Update Jan 2008

Covered key items:
- Project-internal communication of key project info
  - Orientation and best practices workshop dates and locations
  - Preparation for the project
  - Invoicing
  - Participation and confidentiality agreement
  - Recap of confirmed participants
  - Consultant team

File name: WSAA_Project-News-2_Jan2008_AM-Project.pdf

WSAA Benchmarking – Project Article in WSAA Bulletin

Covered key items:
- External communication of project outcome
  - The Best Practices Conference in Sydney
  - Key project findings
  - Example of an inter-regional comparison
  - Comparison by utility characteristics
  - Study tours
  - Communication of results

File name: WSAA_Project-Article_WSAA-Bulletin_2008-12.pdf

H. ORIENTATION AND TRAINING

Orientation and training activities are carried out as soon as the project is started. Kick-off workshops facilitate the transfer and exchange of information necessary to cope with the subsequent performance assessment and performance improvement phases. Recommendations are given in chapter 6.

WSAA Benchmarking – Agenda for Orientation Workshops

Covered key items:
- Welcome and introduction
- Project details
- Participant guide
- Interviews and consultant review
- Aquamark introduction
- Questions and answers
- Wrap up

File name: WSAA_Aquamark-User-Manual_Version3.pdf (page 11ff)

WSAA Benchmarking – Agenda for Accreditation Workshop

Covered key items:
- Objective of the workshop
- Introduction, overview and key principles of benchmarking
- Overview of the assessment framework
- How to use the software?
- Test driving of the assessment
- Overview of the auditing process
- Test driving of project auditing
- How does Aquamark assess our performance?

File name: WSAA_Accreditation-Workshop_Agenda.pdf

I. PROJECT CONTROL?

In close relation with orientation and training activities, documents on project control like guidelines for data acquisition, data validation and handling the assessment results shall be worked out and handed over to the involved parties. Recommendations are given in chapter 6.

WSAA Benchmarking – Aquamark Assessment Guide

Covered key items:
- Objectives and structure of the guidelines
- Guidelines for participants
  - Expectations of utilities
  - Procedures for utilities
- Guidelines for the external review consultant
  - Expectations of reviewers
  - Procedures for external review
- Dispute resolution

File name: WSAA_Aquamark-Assessment-Guide.pdf

WSAA Benchmarking – Aquamark User Manual

Covered key items:
- Project overview
- Overview of Aquamark assessment model
  - Functions and processes
  - Scoring process
  - Asset management profile
- Software operation
  - Administration and introduction
  - Managing an existing data set
  - Undertaking comparisons
- Frequently asked questions
- Glossary

File name: WSAA_Aquamark-User-Manual_Version3.pdf

WSAA Benchmarking – Aquamark Audit Protocol

Covered key items:
- Introduction
- Obligations of participants, auditors and WSAA
- Requirements for auditors
  - Formal qualification
  - Training and development

(Continued)

> ## WSAA Benchmarking – Aquamark Audit Protocol *(Continued)*
>
> - – Professional integrity
> - – Confidentiality and intellectual property
> - Audit procedures
> - Dispute resolution
>
> File name: WSAA_Aquamark-Audit-Protocol_Version2.pdf

J. DATA ACQUISITION AND VALIDATION

Collecting and auditing data is the first step in the performance assessment phase. Document examples given in the previous chapter (assessment guidelines, user manuals, audit protocols) are applied in this step. An example for an onsite interview on data collection and audit is given below.

General recommendations on data acquisition and validation are given in chapter 7.

> ## WSAA Benchmarking – Agenda for Onsite Interview
>
> Covered key items:
> - Executive briefing
> - Consultant review and validation
>   - – Corporate policy and business planning
>   - – Asset capability forward planning
>   - – Asset acquisition
>   - – Asset maintainance
>   - – Asset replacement and rehabilitation
>   - – Business support systems
> - Best Practices Review/Definition
>
> File name: WSAA_Onsite-Interview_Revised-Agenda_Anchorage.pdf

K. ASSESSMENT REPORTING

After data collection and validation and data analysis are completed, performance comparison results can be reported by written reports (draft & final) and by oral presentations at joint best practice workshops or at individual onsite workshops.

General recommendations on assessment reporting are given in chapter 8.

K.1. Utility reports

OVGW Benchmarking – Utility Report

Covered key items:
- Report of a single indicator to a utility
- Definition of the indicator
- Utility's value compared to peers
- Analysis of individual result
- Recommended values

File name: OVGW_utility-individual-report_extract_v2.pdf

WSAA Benchmarking – Draft Utility Report (Anchorage)

Covered key items:
- Executive summary
- Benchmarking participant group
- Comparative benchmarking results
- Business drivers analysis
- Anchorage Water and Wastewater Utility strengths in asset management
- Anchorage Water and Wastewater Utility key opportunities and process gaps
- Anchorage Water and Wastewater Utility initiatives for improvement (9)

File name: WSAA_Utility-Report_Draft_Anchorage.pdf

WSAA Benchmarking – Final Utility Report (Seattle)

Covered key items:
- Executive summary
- Benchmarking participant group
- Comparative benchmarking results
- Business drivers analysis
- Internal comparison to 2004 Aquamark results
- Seattle Public Utilities – Water strengths in asset management
- Seattle Public Utilities – Water key opportunities and process gaps
- Seattle Public Utilities – Water initiatives for improvement (6)

File name: WSAA_Utility-Report_Final_Seattle.pdf

K.2. Public reports

European Benchmarking Cooperation – Project Overview IB08

Covered key items:
- Executive summary
- About EBC
- Project objectives & deliverables
- EBC value proposition
- Benchmarking process & methodology
- Participants
- Reporting
- Geneva workshop
- Evaluation
- To conclude

File name: EBC_Project_Overview_exercise_IB2008.pdf

VEWIN Dutch Benchmarking – Public Report 2006

Covered key items:
- Summary
- Introduction
- Water quality
- Service
- Environment
- Finance & efficiency
- 5 appendices

File name: VEWIN_Public-report_Dutch-benchmarking_2006.pdf

WSAA Benchmarking – Final Project Summary

Covered key items:
- Project overview
- The project
- The participants
- Environmental forces facing the water sector

(Continued)

WSAA Benchmarking – Final Project Summary
(Continued)

- Participant group business drivers
- High level benchmarking results
- Regional comparisons
- Comparison by utility characteristics
- Leading practice themes
- Major improvement initiatives
- Trends in the benchmarking results: from 2004 to 2008
- Concluding remarks

File name: WSAA_Summary-Paper_Asset-Management-2008.pdf

K.3. Best practice workshops

European Benchmarking Cooperation – Agenda for Best Practice Workshop

Covered key items:
- Performance assessment results for drinking water and wastewater
  - Water quality
  - Reliability
  - Service quality
  - Sustainability
  - Finance & efficiency
- EBC's assessment model: objectives, indicators, new benchmarking needs
- Best practices presentations & special topics
- Training session on process improvement

File name: EBC_outline_BP-workshop_exercise_IB2008.pdf

Austrian OVGW Benchmarking – Agenda for Best Practice Workshop on Water Loss Management

Covered key items:
- Performance assessment results
  - Key drivers & scope of WLM
  - WL monitoring, leak detection & strategies
  - Costs & quality results

(Continued)

Austrian OVGW Benchmarking – Agenda for Best Practice Workshop on Water Loss Management
(*Continued*)

- Best practices presentations
  - Rotational leak detection
  - Monitoring & DMAs
  - Noise logging
- Knowledge transfer & group findings

File name: OVGW_Project-Scope_Process-Level.pdf

K.4. Onsite presentations and utility-individual workshops

WSAA Benchmarking – Executive Briefing and Close Out Briefing for Anchorage Water and Wastewater Utility

Covered key items:
- Project Overview
- Self-assessment benchmark comparisons for your utility
- Business driver analysis
- Preliminary findings on sampling and validation
- Areas of difference and areas of consistency
- Selected improvements for discussion and modification

File name: WSAA_Onsite_Presentation_Anchorage.pdf

WSAA Benchmarking – Executive Briefing for City of Tacoma – Water

Covered key items:
- Project Overview
- Self-assessment benchmark comparisons for your utility
- Business driver analysis

File name: WSAA_Onsite_Presentation_v0_Tacoma.pdf

L. IMPROVEMENT ACTIONS

After the performance assessment, the performance improvement phase begins with action planning. In order to prioritize and select possible improvement actions, more analyses are needed for management decisions. An anonymous example is provided.

General recommendations on performance improvement are given in chapter 9 and 10.

> ### Austrian OVGW Benchmarking – Utility Action Planning
>
> Covered key items:
> - List of possible cost saving measures
> - Estimated monetary benefits
> - Different action potential
>   - Action in progress
>   - Influenceable
>   - Not influenceable
>
> File name: OVGW_Project-Scope_Process-Level.pdf

M. PROJECT REVIEW

As pointed out in chapter 10, the final step in a benchmarking exercise is to evaluate the outcomes of the efforts. Feedback from participants will be necessary to

- determine the extent of efforts across utilities in implementing improvements,
- to measure the collective progress against several criteria such as schedule, budget, achievement of service levels, quality of the improvement, as well as impact and performance results,
- to complete documentation on the project by validating and reviewing at a final workshop,
- to collect information on further improvement opportunities for future benchmarking projects,
- to seek input or suggest topics or plans for future benchmarking projects.

Of lesser value than a final workshop, but still a worthwhile effort is to obtain some form of survey information (see example).

European Benchmarking Cooperation – Project Evaluation Form

Covered key items:
- Project organization
- Project phasing
- Drinking water methodology
- Wastewater methodology
- Drinking water report
- Wastewater report
- General

File name: EBC_Evaluation_exercise_IB2008.pdf

Annex C

IWA performance indicators

This annex includes a list of the IWA performance indicators for water supply and wastewater (Alegre *et al.*, 2006 and Matos *et al.*, 2003). The list is merely informative and should not be used to set up a performance assessment system.

For full information on the IWA performance indicators systems, please refer to the corresponding manuals of best practices.

The lists of variables and context information elements from the IWA systems can also be downloaded from the website of the Specialist Group on Benchmarking and Performance Assessment (see Annex B for more information):

- PI-list_water-supply-manual.pdf
- PI-list_waste-water-manual.pdf

A. PERFORMANCE INDICATORS FOR WATER SUPPLY SERVICES

Water resources indicators (page II-31)

WR1 – Inefficiency of use of water resources (%)
WR2 – Water resources availability (%)
WR3 – Own water resources availability (%)
WR4 – Reused supplied water (%)

Personnel indicators (page II-33)	
Total personnel (page II-33)	
Pe1 –	Employees per connection (No./1000 connections)
Pe2 –	Employees per water produced (No./(10^6 m^3/year))
Personnel per main function (page II-33)	
Pe3 –	General management personnel (%)
Pe4 –	Human resources management personnel (%)
Pe5 –	Financial and commercial personnel (%)
Pe6 –	Customer service personnel (%)
Pe7 –	Technical services personnel (%)
	Pe8 – Planning & construction personnel (%)
	Pe9 – Operations & maintenance personnel (%)
Technical services personnel per activity (page II-35)	
Pe10 –	Water resources and catchment management personnel (No./(10^6 m^3/year))
Pe11 –	Abstraction and treatment personnel (No./(10^6 m^3/year))
Pe12 –	Transmission, storage and distribution personnel (No./100 km)
Pe13 –	Water quality monitoring personnel (No./(10000 tests/year))
Pe14 –	Meter management personnel (No./1000 meters)
Pe15 –	Support services personnel (%)
Personnel qualification (page II-36)	
Pe16 –	University degree personnel (%)
Pe17 –	Basic education personnel (%)
Pe18 –	Other qualification personnel (%)
Personnel training (page II-37)	
Pe19 –	Total training (hours/employee/year)
	Pe20 – Internal training (hours/employee/year)
	Pe21 – External training (hours/employee/year)

Personnel health and safety (page II-38)	
Pe22 –	Working accidents (No./100 employees/year)
Pe23 –	Absenteeism (days/employee/year)
	Pe24 – Absenteeism due to working accidents or illness at work (days/employee/year)
	Pe25 – Absenteeism due to other reasons (days/employee/year)
Overtime work (page II-39)	
Pe26 –	Overtime work (%)

Physical indicators (page II-39)	
Treatment (page II-39)	
Ph1 –	Treatment plant utilisation (%)

Storage (page II-40)	
Ph2 –	Raw water storage capacity (days)
Ph3 –	Transmission and distribution storage capacity (days)
Pumping (page II-40)	
Ph4 –	Pumping utilisation (%)
Ph5 –	Standardised energy consumption ($kWh/m^3/100\,m$)
Ph6 –	Reactive energy consumption (%)
Ph7 –	Energy recovery (%)
Transmission and distribution (page II-41)	
Ph8 –	Valve density (No./km)
Ph9 –	Hydrant density (No./km)
Meters (page II-42)	
	Ph10 – District meter density (No./1000 service connections)
	Ph11 – Customer meter density (No./service connection)
	Ph12 – Metered customers (No./customer)
	Ph13 – Metered residential customers (No./customer)
Automation and control (page II-43)	
Ph14 –	Automation degree (%)
Ph15 –	Remote control degree (%)

Operational indicators (page II-43)	
Inspection and maintenance of physical assets (page II-43)	
Op1 –	Pump inspection (–/year)
Op2 –	Storage tank cleaning (–/year)
Op3 –	Network inspection (%/year)
Op4 –	Leakage control (%/year)
Op5 –	Active leakage control repairs (No./100 km/year)
Op6 –	Hydrant inspection (–/year)
Instrumentation calibration (page II-45)	
Op7 –	System flow meters calibration (–/year)
Op8 –	Meter replacement (–/year)
Op9 –	Pressure meters calibration (–/year)
Op10 –	Water level meters calibration (–/year)
Op11 –	On-line water quality monitoring equipment calibration (–/year)
Electrical and signal transmission equipment inspection (page II-47)	
Op12 –	Emergency power system inspection (–/year)
Op13 –	Signal transmission equipment inspection (–/year)
Op14 –	Electrical switchgear equipment inspection (–/year)
Vehicle availability (page II-48)	
Op15 –	Vehicle availability (No./100 km)
Mains, valves and service connection rehabilitation (page II-48)	
Op16 –	Mains rehabilitation (%/year)
	Op17 – Mains renovation (%/year)
	Op18 – Mains replacement (%/year)
	Op19 – Replaced valves (%/year)
Op20 –	Service connection rehabilitation (%/year)
Pumps rehabilitation (page II-50)	
	Op21 – Pump refurbishment (%/year)
	Op22 – Pump replacement (%/year)

Operational water losses (page II-50)	
Op23 –	Water losses per connection (m^3/connection/year)
Op24 –	Water losses per mains length (m^3/km/day)
	Op25 – Apparent losses (%)
	Op26 – Apparent losses per system input volume (%)
	Op27 – Real losses per connection (l/connection/day when system is pressurised)
	Op28 – Real losses per mains length (l/km/day when system is pressurised)
Op29 –	Infrastructure leakage index (–)
Failure (page II-53)	
Op30 –	Pump failures (days/pump/year)
Op31 –	Mains failures (No./100 km/year)
Op32 –	Service connection failures (No./1000 connections/year)
Op33 –	Hydrant failures (No./1000 hydrants/year)
Op34 –	Power failures (hours/pumping station/year)
Water metering (page II-56)	
Op36 –	Customer reading efficiency (–)
Op37 –	Residential customer reading efficiency (–)
Op38 –	Operational meters (%)
Op39 –	Unmetered water (%)
Water quality monitoring (page II-57)	
Op40 –	Tests carried out (%)
	Op41 – Aesthetic tests carried out (%)
	Op42 – Microbiological tests carried out (%)
	Op43 – Physical-chemical tests carried out (%)
	Op44 – Radioactivity tests carried out (%)

Quality of service indicators (page II-59)	
Service coverage (page II-59)	
QS1 –	Households and businesses supply coverage (%)
QS2 –	Buildings supply coverage (%)
QS3 –	Population coverage (%)
	QS4 – Population coverage with service connections (%)
	QS5 – Population coverage with public taps or standpipes (%)
Public taps and standpipes (page II-61)	
QS6 –	Operational water points (%)
QS7 –	Average distance from waterpoints to households (m)
QS8 –	Per capita water consumed in public taps and standpipes (l/person/day)
QS9 –	Population per public tap or standpipe (persons/tap)
Pressure and continuity of supply (page II-62)	
QS10 –	Pressure of supply adequacy (%)
QS11 –	Bulk supply adequacy (%)
QS12 –	Continuity of supply (%)
QS13 –	Water interruptions (%)
QS14 –	Interruptions per connection (No./1000 connections/year)
QS15 –	Bulk supply interruptions (No./ delivery point /year)
QS16 –	Population experiencing restrictions to water service (%)
QS17 –	Days with restrictions to water service (%)
Quality of supplied water (page II-65)	
QS18 –	Quality of supplied water (%)
	QS19 – Aesthetic tests compliance (%)
	QS20 – Microbiological tests compliance (%)
	QS21 – Physical-chemical tests compliance (%)
	QS22 – Radioactivity tests compliance (%)
Service connection and meter installation and repair (page II-66)	
QS23 –	New connection efficiency (days)
QS24 –	Time to install a customer meter (days)
QS25 –	Connection repair time (days)

Customer complaints (page II-67)	
QS26 –	Service complaints per connection (No. complaints/1000 connections/year)
QS27 –	Service complaints per customer (No. complaints/customer/year)
	QS28 – Pressure complaints (%)
	QS29 – Continuity complaints (%)
	QS30 – Water quality complaints (%)
	QS31 – Interruption complaints (%)
QS32 –	Billing complaints and queries (No./customer/year)
QS33 –	Other complaints and queries (No./customer/year)
QS34 –	Response to written complaints (%)

Economic and financial indicators (page II-70)	
Revenues (page II-70)	
Fi1 –	Unit revenue (EUR/m^3)
	Fi2 – Sales revenues (%)
	Fi3 – Other revenues (%)
Costs (page II-71)	
Fi4 –	Unit total costs (EUR/m^3)
	Fi5 – Unit running costs (EUR/m^3)
	Fi6 – Unit capital costs (EUR/m^3)
Composition of running costs per type of costs (page II-71)	
	Fi7 – Internal manpower costs (%)
	Fi8 – External services costs (%)
	Fi9 – Imported (raw and treated) water costs (%)
	Fi10 – Electrical energy costs (%)
	Fi11 – Other costs (%)
Composition of running costs per main function of the water undertaking (page II-73)	
	Fi12 – General management functions costs (%)
	Fi13 – Human resources management functions costs (%)
	Fi14 – Financial and commercial functions costs (%)
	Fi15 – Customer service functions costs (%)
	Fi16 – Technical services functions costs (%)

Composition of running costs per technical function activity (page II-74)		
		Fi17 – Water resources and catchment management costs (%)
		Fi18 – Abstraction and treatment costs (%)
		Fi19 – Transmission, storage and distribution costs (%)
		Fi20 – Water quality monitoring costs (%)
		Fi21 – Meter management costs (%)
		Fi22 – Support services costs (%)
Composition of capital costs (page II-75)		
		Fi23 – Depreciation costs (%)
		Fi24 – Net interest costs (%)
Investment (page II-76)		
Fi25 –	Unit investment (EUR/m^3)	
		Fi26 – Investments for new assets and reinforcement of existing assets (%)
		Fi27 – Investments for asset replacement and renovation (%)
Average water charges (page II-77)		
Fi28 –	Average water charges for direct consumption (EUR/m^3)	
Fi29 –	Average water charges for exported water (EUR/m^3)	
Efficiency (page II-77)		
Fi30 –	Total cost coverage ratio (–)	
Fi31 –	Operating cost coverage ratio (–)	
Fi32 –	Delay in accounts receivable (days equivalent)	
Fi33 –	Investment ratio (–)	
Fi34 –	Contribution of internal sources to investment = CTI (%)	
Fi35 –	Average age of tangible assets (%)	
Fi36 –	Average depreciation ratio (–)	
Fi37 –	Late payments ratio (–)	
Fi38 –	Inventory value (–)	
Leverage (page II-79)		
Fi39 –	Debt service coverage ratio = DSC (%)	
Fi40 –	Debt equity ratio (–)	
Liquidity (page II-80)		
Fi41 –	Current ratio (–)	

Profitability (page II-80)	
Fi42	Return on net fixed assets (%)
Fi43	Return on equity (%)
Fi44	Return on capital employed (%)
Fi45	Asset turnover ratio (−)
Economic water losses (page II-81)	
Fi46 −	Non-revenue water by volume (%)
Fi47 −	Non-revenue water by cost (%)

B. PERFORMANCE INDICATORS FOR WASTE WATER SERVICES

Environmental indicators (wEn)	
Wastewater	
wEn1	WWTP compliance with discharge consents
wEn2	Wastewater reuse
wEn3	Intermittent overflow discharge frequency
wEn4	Intermittent overflow discharge volume
wEn5	Intermittent overflow discharge related to rainfall
Solid residues	
wEn6	Sludge production in WWTP
wEn7	Sludge utilisation
wEn8	Sludge disposed
wEn9	− sludge going to landfill
wEn10	− sludge thermally processed
wEn11	− other sludge disposal
wEn12	Sediments from sewers
wEn13	Sediments from ancillaries
wEn14	Solid waste from screens and grit
wEn15	Sediments from on-site systems

Personnel indicators (wPe)	
Total personnel	
wPe1	Personnel in WWT per population equivalent
wPe2	Personnel working on SE per length of sewer
Personnel per main function	
wPe3	General management personnel
wPe4	Human resources management personnel
wPe5	Financial and commercial services personnel
wPe6	Customer service personnel
wPe7	Technical services personnel
wPe8	– planning, design and construction personnel
wPe9	– operation and maintenance personnel
Technical personnel per activity	
wPe10	Technical WWT personnel
wPe11	Technical SE personnel
wPe12	Wastewater quality monitoring personnel
wPe13	Support services personnel
wPe14	University degree personnel
Personnel qualification	
wPe15	Basic education personnel
wPe16	Other qualification personnel
Personnel training	
wPe17	Total training of personnel
Personnel vaccination and safety	
wPe18	Vaccination
wPe19	Authorised confined spaces training
wPe20	Working accidents
wPe21	– personnel working fatalities
Absenteeism	
wPe22	Absenteeism
wPe23	– absenteeism due to accidents or illness at work
wPe24	– absenteeism due to other reasons
Overtime work	
wPe25	Overtime work

Physical indicators (wPh)	
Wastewater treatment	
wPh1	Preliminary treatment utilisation
wPh2	Primary treatment utilisation
wPh3	Secondary treatment utilisation
wPh4	Tertiary treatment utilisation
Sewers	
wPh5	Surcharging in gravity sewers in dry weather
wPh6	Surcharging in gravity sewers in wet weather
wPh7	High sewer surcharging
Pumping headroom	
wPh8	Pump power utilised in SE
wPh9	Pump power utilised in WWTP
wPh10	Sewer system pump headroom
Automation and control	
wPh11	Automation degree
wPh12	Remote control degree
Operational indicators (wOp)	
Sewer inspection and maintenance	
wOp1	Sewer cleaning
wOp2	Sewer inspection
wOp3	Manhole inspection
wOp4	Gully pot cleaning
wOp5	Gully pot inspection
Tanks and csos inspection and maintenance	
wOp6	Tank and CSO inspection frequency
wOp7	Tank and CSO inspection volume
wOp8	Tank and CSO cleaning
wOp9	Screen inspection
Pumps and pumping stations inspection	
wOp10	Pumping station inspection frequency
wOp11	Pump inspection by power

Equipment calibration	
wOp12	SE flow meters calibration
wOp13	WWTP flow meters calibration
wOp14	Wastewater quality monitoring equipment
Electrical and signal transmission equipment inspection	
wOp15	Emergency power system inspection
wOp16	Signal transmission equipment inspection
wOp17	Electrical switchgear equipment inspection
Energy consumption	
wOp18	Energy consumption at WWT
wOp19	WWT energy recovery from co-generation processes
wOp20	Standardised energy consumption
Sewer system rehabilitation	
wOp21	Sewer rehabilitation
wOp22	– Sewers renovation
wOp23	– Sewers replacement
wOp24	– Sewers and joints repair
wOp25	Manhole chambers replacement, renewal, renovation or repair
wOp26	Replacement of manhole covers
wOp27	Service connection rehabilitation
Pumps rehabilitation	
wOp28	Pump refurbishment
wOp29	Pump replacement
Inflow/infiltration/exfiltration (I/I/E)	
wOp30	Inflow/Infiltration/Exfiltration (I/I/E)
wOp31	Inflow
wOp32	Infiltration
wOp33	Exfiltration
Failures	
wOp34	Sewer blockages
wOp35	– sewer blockage locations
wOp36	Pumping station blockages

wOp37	Flooding from sanitary sewers
wOp38	Flooding from combined sewers
wOp39	Surface flooding
wOp40	Sewer collapses
wOp41	Pump failures
wOp42	Power failures
CSO control	
wOp43	CSO control
Wastewater and sludge quality monitoring	
wOp44	Wastewater quality tests carried out
wOp45	– BOD tests
wOp46	– COD tests
wOp47	– TSS tests
wOp48	– total phosphorus tests
wOp49	– nitrogen tests
wOp50	– faecal E.coli tests
wOp51	– other tests
wOp52	Sludge tests carried out
wOp53	Industrial discharges tests carried out
Vehicle availability	
wOp54	Vehicle availability
Safety equipment	
wOp55	Gas detectors
wOp56	Permanently installed gas detectors
Quality of service indicators (wQS)	
Population served	
wQS1	Resident population connected to sewer system
wQS2	Resident population served by WWTP
wQS3	Resident population served by on-site systems
wQS4	Resident population not served
Treated wastewater	
wQS5	Treated wastewater in WWTP
wQS6	– preliminary treatment

wQS7	– primary treatment
wQS8	– secondary treatment
wQS9	– tertiary treatment
Flooding	
wQS10	Flooding of properties from sanitary sewers in dry weather
wQS11	Flooding of properties from sanitary sewers in wet weather
wQS12	Flooding of properties from combined sewers in dry weather
wQS13	Flooding of properties from combined sewers in wet weather
wQS14	Surface water flooding of properties in wet weather
Interruptions	
wQS15	Interruption of wastewater collection and transport services
Reply to customer requests	
wQS16	New service connection efficiency
wQS17	Service connection repair time
wQS18	Average response time to empty septic tanks or pits
Complaints	
wQS19	Total complaints
wQS20	– blockage complaints
wQS21	– flooding complaints
wQS22	– pollution incidents complaints
wQS23	– odour complaints
wQS24	– rodents related complaints
wQS25	– customer account related complaints
wQS26	– other complaints
wQS27	Response to complaints
Third party damages	
wQS28	Responsibility for third party damages
Impacts on traffic	
wQS29	Traffic disturbances
Economic and financial indicators (wFi)	
Revenues	
wFi1	Unit revenue
wFi2	– service revenues

wFi3	– other revenues
wFi4	Industrial revenues
Costs	
wFi5	Unit total cost per p.e.
wFi6	Unit total cost per length of sewer
wFi7	– unit running cost per p.e.
wFi8	– unit running cost per length of sewer
wFi9	– unit capital cost per p.e.
wFi10	– unit capital cost per length of sewer
Composition of running costs per type of costs	
wFi11	Internal manpower costs
wFi12	External services costs
wFi13	Energy costs
wFi14	Materials, chemicals and other consumable costs
wFi15	Other running costs
Composition of running costs per main function (internal and outsourced)	
wFi16	General management functions costs
wFi17	Human resources management functions costs
wFi18	Financial and commercial management functions costs
wFi19	Customer service management costs
wFi20	Technical services functions costs
Composition of running costs per technical function activity	
wFi21	Wastewater treatment running costs
wFi22	Sewer system running costs
wFi23	Wastewater quality monitoring running costs
wFi24	Support services running costs
Composition of capital costs	
wFi25	Depreciation costs
wFi26	Net interest costs
Investment	
wFi27	Unit investment
wFi28	– investments for new assets and reinforcement of existing assets
wFi29	– investments for asset replacement and renovation

Efficiency indicators	
wFi30	Total cost coverage ratio
wFi31	Operating cost coverage ratio
wFi32	Delay in accounts receivable
wFi33	Investment ratio
wFi34	Contribution of internal sources to investment = CTI
wFi35	Average age of tangible assets
wFi36	Average depreciation ratio
wFi37	Late payments ratio
wFi38	Inventory value
Leverage indicators	
wFi39	Debt service coverage ratio = DSC
wFi40	Debt equity ratio
Liquidity indicator	
wFi41	Current ratio
Profitability indicators	
wFi42	Return on net fixed assets
wFi43	Return on equity
wFi44	Return on capital employed
wFi45	Asset turnover ratio

References and select bibliography

Alegre H., Baptista J. M., Cabrera Jr, E., Cubillo F., Duarte P., Hirner W., Merkel W. and Parena R. (2006). *Performance Indicators for Water Supply Services. Manual of Best Practice*. 2nd edition, IWA Publishing, London, UK.

Brueck T. (2005). *Developing and Implementing a Performance Measurement System: Volume II*. WERF and AwwaRF.

Cabrera Jr, E., Alegre H. and Merkel W. (2009b). Performance Assessment of Urban Utilities: the Case of Water Supply, Wastewater and Solid Water. *Journal of Water Supply: Research and Technology*, AQUA, **58**(5), 305–315.

Cabrera Jr, E., Dane P., Haskins Sc. and Theuretzbacher-Fritz H. (2009a). Benchmarking Revisited: The Global Search for Performance Improvement. *Journal/American Water Works Association*, JOURNAL AWWA, **101**(11), 22–26.

Cabrera Jr, E. and Pardo M. A. (Ed) (2008). *COST C18 – Performance Assessment of Urban Infrastructure Services. Drinking Water, Wastewater and Solid Waste*. Proceedings of COST C18 Final Conference & IWA Conference PI08, Valencia, IWA Publishing, London, UK.

Camp R. (1989). *Benchmarking: the Search for Industry Best Practices that Lead to Superior Performance*. Productivity press, New York, USA.

Camp R. (1998). *Global Cases in Benchmarking: Best Practices from Organizations Around the World*. ASQ Quality press, Milwaukee, USA.

Carlson S. and Walburger A. (2007). *Energy Index Development for Benchmarking Water and Wastewater Utilities*. Awwa Research Foundation.

ISO 24510:2007. *Activities Relating to Drinking Water and Wastewater Services – Guidelines for the Assessment and for the Improvement of the Service to Users*.

ISO 24511:2007. *Activities Relating to Drinking Water and Wastewater Services – Guidelines for the Management of Wastewater Utilities and for the Assessment of Wastewater Services*.

ISO 24512:2007. *Activities Relating to Drinking Water and Wastewater Services – Guidelines for the Management of Drinking Water Utilities and for the Assessment of Drinking Water Services.*

IWA, VEWIN, WATERNETWERK (2009). *PI09 Benchmarking Water Services – the way forward.* Proceedings of the IWA PI09 Conference, Amsterdam, The Netherlands, CD ROM.

Kaplan R. S. and Norton D. P. (1996). *The Balanced Scorecard Translating Strategy into Action.* HBS Press.

Kenway S., Howe C. and Maheepala S. (2008). *Triple Bottom Line Reporting of Sustainable Water Utility Performance.* AwwaRF.

Kingdom B., Knapp J., LaChange P. and Olstein M. (1996). *Performance Benchmarking for Water Utilities.* AWWA, Denver, USA.

Koelbl J. (2009). Process Benchmarking in the Water Supply Sector: Management of Physical Water Losses. PhD thesis, Graz University of Technology, Austria, ISBN 978-3-85125-055-8.

Koelbl J., Mayr E., Theuretzbacher-Fritz H., Neunteufel R. and Perfler R. (2009). *Is it Possible to Benchmark the Process of Physical Water Loss Management?* In: Proceedings of IWA conference "PI09: Benchmarking water services – the way forward". Amsterdam, The Netherlands.

Lambert A. and Hirner W. (2000). Losses from Water Supply Systems: Standard Terminology and Recommended Performance Measures. *IWA blue pages*, International Water Association, 13 p.

Larsson M., Parena R., Smeets E. and Troquet I. (2002). *Process Benchmarking in the Water Industry – towards a worldwide approach.* IWA Publishing, London, UK.

Matos R., Cardoso R., Ashley R., Duarte P., Molinari A. and Schulz A. (2003). *Performance Indicators for Wastewater Services. Manual of Best Practice.* IWA Publishing, London, UK.

Neunteufel R., Theuretzbacher-Fritz H., Koelbl J., Perfler R., Friedl F. and Mayr E. (2009). *Benchmarking and Best Practices within the Austrian Water Supply.* Public report on OVGW Benchmarking 2008 (Stage C), Vienna-Graz, Austria.

Stedman L. (2007). Benchmarking best practice in the water sector. *Water Utility Management International*, **2**(3), 20–23. IWA Publishing, London, UK.

Theuretzbacher-Fritz H., Koelbl J., Neunteufel R., Mayr E., Friedl F. and Perfler R. (2009). *Are Unit Costs Comparable Regarding Efficiency?* Proceedings of IWA conference "PI09: Benchmarking water services – the way forward". Amsterdam, The Netherlands.

VEWIN. (2007). *Reflections on Performance 2006.* Public report on Benchmarking in the Dutch drinking water industry.

VEWIN. (2010). *Reflections on Performance 2009.* Public report on Benchmarking in the Dutch drinking water industry.

Water Research Foundation. (2009). *Enhancement of QualServe Tools to Improve Utility Operations.* AWWA

Water Utility Management International. (2008). Special issue Benchmarking. *Water Utility Management International*, **3**(2), 32 p. IWA Publishing, London, UK.

Zelig N. (Ed.) (2003). *The Evolving Water Utility: Pathways to Higher Performance.* AWWA.

IWA Publishing's authorised EU representative for General Product
Safety Regulations is Diane D'Arras, 15 rue Duret, 75116 Paris,
France, e-mail: safety@iwap.co.uk.

Printed and bound by CPI Group (UK) Ltd, Croydon, CR0 4YY

12/05/2026

02108880-0009